An Illustrated History of L.M.S. LOCOMOTIVES

Frontispiece Peter Drummond contrasts in Scotland: ex-HR 'Ben' class 4-4-0 No. 14406 *Ben Slioch* and ex-GSWR 16 class 2-6-0 No. 17829 typify this controversial man's early and late work on the two smaller Scottish constituents of the LMS.

Stephen Collection, courtesy NRM and Photomatic

An Illustrated History of L.M.S. Locomotives

Volume Three: Absorbed Pre-Group Classes, Northern Division

by Bob Essery & David Jenkinson

Silver Link Publishing Ltd

© R. J. Essery and D. Jenkinson 1986 and 1994

All rights reserved. No part of this publication may be reproduced, stored in a retrieval system or transmitted, in any form or by any means, electronic, mechanical, photocopying, recording or otherwise, without prior permission in writing from Silver Link Publishing Ltd.

First published by Oxford Publishing Co 1986
Silver Link Publishing Ltd edition first published May 1994

British Library Cataloguing in Publication Data

A catalogue record for this book is available from the British Library

ISBN 1 85794 024 5

Silver Link Publishing Ltd
Unit 5
Home Farm Close
Church Street
Wadenhoe
Peterborough PE8 5TE
Tel/fax (0832) 720440

Printed and bound in Great Britain

Title Page Ex-CR Drummond/Lambie 4-4-0 No. 14309. This was a 'composite' rebuild in 1907 - *see page 13*.
Authors' Collection

The fine lines of the ex-CR 'Dunalastair II' class are well displayed in this pre-1928 view of No. 14336, the last of the saturated batch — *see page 16*.
Stephen Collection, courtesy NRM

Contents

		Page
	Authors' Introduction to Volume Three	vii
	Bibliographical Note	vii
	Locomotive Livery Key List and notes on its use	viii
	Introduction to the Northern Division	1
Chapter 1	Caledonian Railway — Introduction and Passenger Tender Classes	3
Chapter 2	Caledonian Railway — Passenger Tank Classes	45
Chapter 3	Caledonian Railway — Freight Tank Classes	65
Chapter 4	Caledonian Railway — Freight Tender Classes	85
Chapter 5	Glasgow and South Western Railway — Introduction and Passenger Classes	107
Chapter 6	Glasgow and South Western Railway — Freight Classes	139
Chapter 7	Highland Railway — All Classes	169
Chapter 8	The 'Drummond' factor on the LMS in Scotland	209
Appendix	Corrections and Additions to Volumes One and Two	215
	Index of Locomotive Classes (Capital Stock) covered in Volume Three	215

This view of 'River' class 4-6-0 No. CR943, later LMS No. 14761 *(see page 37)* at Carlisle, Canal Junction, with a very 'mixed bag' of vehicles behind its tender nicely symbolises the Northern Division at the start of the LMS period. Designed for the HR, the engines were bought by the CR and went back to the Highlands after the Grouping.

Bernard Matthews Collection

Caledonian Contrasts These two views showing Brittain 670 class 0-4-2 No. 17018 and McIntosh 812 class 0-6-0 No. 17611 effectively symbolise the transition already wrought at St. Rollox on the Caledonian Railway in pre-LMS days. About the only thing in common is the plain black colour scheme and the chimney shape — pure Drummond in fact!
H.C. Casserley and Authors' Collection

Authors' Introduction to Volume Three

It was, we suppose, somewhat inevitable, having been given our publisher's permission to expand this series from its original three volume concept — *see Preface to Volume Two* — that matters should have got even further out of hand! So, once again, we start with an apology, this time for the fact that this book in fact contains only the Northern Division (Scottish) pre-group locomotives of the LMS and not the expected Midland Division fleet as well. Now essentially, the object of our original decision to increase the number of volumes was to make each part reasonably manageable and not too prohibitive, economically; but it fairly soon became clear that this was not going to be achieved with *Volume Three* as originally planned.

Basically, our fundamental error was to assume that the allocation of space between *Volumes Two and Three* could be done on a simple 'number count' of the engines involved. What we should have done was to reckon up the number of different classes concerned. On this basis, the Northern Division generated almost as many different types as the Western and Central Divisions combined. Had we added to these all the Midland engines as well, not to mention the LMS types built to MR design, the picture count would have risen to some 800, the page count to around 400 plus, and the price tag somewhere into the stratospheric levels! Therefore, and with the publisher's blessing, rather than offer the reader a 'jumbo' sized *Volume Three*, we have opted for two rather smaller compilations, of which this is the first. Hopefully, it will be followed, quite quickly, by the Midland Division *Volume Four* which, even at the time of going to press with this book, is some 90 per cent complete.

Some readers may express surprise that we have, apparently, relegated the Midland engines almost to the last place in the pre-group survey — but this is deliberate. After the LNWR fleet *(Volume Two)*, the Midland locomotives were the next largest numerical group of engines to be absorbed by the LMS, but were far more influential in post-group LMS thinking. We have, therefore, chosen to place them immediately before the study of the LMS standard types now to be covered in *Volume Five*, thus maintaining an element of continuity in the story.

As with *Volume Two*, so too in this book we shall continue to place most of the emphasis on the external style(s) of LMS locomotives, leaving the overtly technical matters mostly to one side, except where it seems necessary to cover them in order to explain visual differences. We shall not endeavour to cross every 't' or dot every 'i' and, doubtless, the ultra pedantic will find points on which to take issue. If they find genuine mistakes we hope that they will inform us via the publisher, but, as we said in the introduction to *Volume Two*, our main objective is to try and bring a little order out of chaos to help readers, especially those of the newer generation, to understand a little more of what was, undoubtedly, a complex story. If we achieve no more than this we shall feel well-pleased.

The photographs in this book are the usual mix of official and private views, duly acknowledged wherever we can find evidence of ownership. We would like to thank all the photographers, past and present, known and unknown, whose work has been examined during our researches, and if there are any mistaken attributions, we would welcome being advised via the publisher. Finally, to our friends who have continued to read and monitor the work, we offer a particularly special vote of thanks for their help and encouragement.

RJE
Solihull
1986

DJ
Knaresborough
1986

BIBLIOGRAPHICAL NOTE

In *Volume One* we gave an extended bibliography of sources, and we shall try to bring this up to date in the last volume of the series. However, for readers who wish to have one or two particularly relevant books to hand while reading this part of our survey, we would strongly recommend the fairly recent multi-part compilation under the series title *British Locomotive Catalogue, 1825-1923* published by Moorland Publishing Company of Ashbourne, Derbyshire. The three volumes of this eight-part series relevant to the LMS are *Volumes 2, 3 and 4* and probably give all the building, numbering and other data which any reader could reasonably want. They are not illustrated but they complement (and add to) the Casserley/Johnson *Locomotives at the Grouping* to which we referred in *Volume One* as a valuable stand-by.

In preparing this book, we have also found the following titles particularly helpful:

Forty Years of Caledonian Locomotives — H. J. Campbell Cornwell, David & Charles, 1974
Locomotives of the Glasgow and South Western Railway — David L. Smith, David & Charles, 1976
A History of Highland Locomotives — Peter Tatlow, OPC, 1979

Locomotive Livery Key List and Note on its use

We repeat, below, *Table 10 from page 204 of Volume One* in order to help the reader. The list should adequately define the vast majority of liveries carried by the LMS locomotives from 1923-47 and is, in essence, a tabulated summary of the livery section of *Volume One*. In virtually all the photographs in this volume, liveries are described by reference to the code letters and numbers in the left-hand column of this list, augmented by such other detail as seems relevant in context. The main variable (apart from totally non-standard paint schemes) was in the centre-to-centre spacing of the letters 'LMS' during the 1928-47 period. Where a particular class of engine displayed some consistency in this respect, we shall say so in the narrative, but where much variety was evident, we shall either try to provide some sort of valid generalisations where we can, or simply draw attention to the problem.

Crimson Lake Livery variations

A1	Pre-1928 standard, 18in. figures,	LMS Coat of Arms
A2	Pre-1928 standard, 18in. figures,	Individual Letters 'LMS'
A3	Pre-1928 standard, 14in. figures,	LMS Coat of Arms
A4	Pre-1928 standard, 14in. figures,	Individual Letters 'LMS'
A5	Post-1927 standard, Gold/Black insignia,	10in. numerals
A6	Post-1927 standard, Gold/Black insignia,	12in. numerals
A7	Post-1927 standard, Gold/Black insignia,	14in. numerals (Midland pattern)
A8	Post-1927 standard, Straw/Black insignia,	10in. numerals
A9	Post-1927 standard, Straw/Black insignia,	12in. numerals
A10	Post-1927 standard, Straw/Black insignia,	14in. numerals (Standard pattern)
A11	Post-1927 standard, Gold/Red insignia,	12in. numerals
A12	Post-1927 standard, Gold/Red insignia,	1936 pattern
A13	Post-1927 standard, Yellow/Red insignia,	10in. numerals
A14	Post-1927 standard, Yellow/Red insignia,	12in. numerals
A15	Post-1927 standard, Yellow/Red insignia,	14in. numerals (Midland pattern)

Lined Black Livery variations

B1	Lined Black livery, Horwich/St. Rollox style,	18in. Midland figures
B2	Post-1927 standard, Gold/Red insignia,	10in. numerals
B3	Post-1927 standard, Gold/Red insignia,	12in. numerals
B4	Post-1927 standard, Gold/Red insignia,	14in. numerals (Midland pattern)
B5	Post-1927 standard, Gold/Black insignia,	10in. numerals
B6	Post-1927 standard, Gold/Black insignia,	12in. numerals
B7	Post-1927 standard, Gold/Black insignia,	14in. numerals (Midland pattern)
B8	Post-1927 standard, Yellow/Red insignia,	10in. numerals
B9	Post-1927 standard, Yellow/Red insignia,	12in. numerals
B10	Post-1927 standard, Yellow/Red insignia,	14in. numerals (Midland pattern)
B11	Post-1927 standard, Gold/Red insignia,	1936 pattern
B12	1946 standard livery — full lining style	
B13	1946 standard livery — simpler original lining style	

Plain Black Livery variations

C1	Pre-1928 standard, 18in. figures,	Standard cab/bunker panel
C2	Pre-1928 standard, 18in. figures,	Round cornered cab/bunker panel
C3	Pre-1928 standard, 18in. figures,	Individual Letters 'LMS'
C4	Pre-1928 standard, 14in. figures,	Standard cab/bunker panel
C5	Pre-1928 standard, 14in. figures,	Round cornered cab/bunker panel
C6	Pre-1928 standard, 14in. figures,	Individual Letters 'LMS'
C7	Crewe 'hybrid' style, 18in. figures,	LMS Coat of Arms
C8	Crewe 'hybrid' style, 14in. figures,	(Midland pattern), LMS Coat of Arms
C9	Crewe 'hybrid' style, 14in. figures,	(Standard pattern — straw), LMS Coat of Arms
C10	Crewe 'hybrid' style, 18in. figures,	Individual Letters 'LMS'
C11	Crewe 'hybrid' style, 14in. figures,	(Midland pattern), Individual Letters 'LMS'
C12	Crewe 'hybrid' style, 14in. figures,	(Standard pattern), Individual Letters 'LMS'
C13	Post-1927 standard, Gold/Black insignia,	10in. numerals
C14	Post-1927 standard, Gold/Black insignia,	12in. numerals
C15	Post-1927 standard, Gold/Black insignia,	14in. numerals (Midland pattern)
C16	Post-1927 standard, Plain Straw insignia,	10in. numerals
C17	Post-1927 standard, Plain Straw insignia,	12in. numerals
C18	Post-1927 standard, Plain Straw insignia,	14in. numerals (Standard pattern)
C19	Post-1927 standard, Gold/Red insignia,	1936 pattern
C20	Post-1927 standard, Gold/Black insignia,	1936 pattern
C21	Post-1927 standard, Yellow/Red insignia,	10in. numerals
C22	Post-1927 standard, Yellow/Red insignia,	12in. numerals
C23	Post-1927 standard, Yellow/Red insignia,	14in. numerals (Midland pattern)
C24	Post-1927 standard, Plain Yellow insignia,	10in. numerals
C25	Post-1927 standard, Plain Yellow insignia,	12in. numerals
C26	Post-1927 standard, Plain Yellow insignia,	14in. numerals (Midland pattern)
C27	1946 standard insignia — smaller size	
C28	1946 standard insignia — larger size	

Introduction to the Northern Division

The Northern Division of the LMS was composed of its three Scottish constituents, the Caledonian, Glasgow and South Western, and the Highland railways. The initial post-grouping LMS number allocation was from 14000-17999, although the highest number actually taken up was 17997 (ex-CR 0-8-0 tender type). This overall series was, as usual, broken down into the normal LMS four-fold category of locomotive types (passenger tender and tank, freight tank and tender) and within each category, ascending power class (by wheel arrangement) was adopted.

This logical system had the effect of putting all engines of one specific class into a continuous number series (except for odd mistakes), but it did cause the three principal company fleets to be intermixed within each power class and/or wheel arrangement block. The general principle was, for any single wheel arrangement, to take Caledonian, G&SW and Highland engines of any one power class in that order before going to the next higher power class, viz:

> Caledonian power class 1
> G&SW power class 1
> Highland power class 1   } for any single wheel arrangement
> Caledonian power class 2
> G&SW power class 2

. . . etc., followed by the next wheel arrangement.

This could have the effect of 'separating' related company classes (e.g. the 4-4-0 types — *see Chapter 1*) if a later development was in a higher power class than its predecessor. This was much more apparent on the Northern Division than on the Midland and Western divisions, or even the Central Division wherein the Furness Railway types were put amongst the LYR engines — *see Volume Two*.

In this survey, we shall, of course, take each company separately and deal with them generally in the same order as did the LMS at the time of the 1923 renumbering. As with the Western and Central divisions *(Volume Two)*, Northern Division engines were given smokebox numberplates pre-1928 but generally had them removed thereafter. It was somewhat unusual to see them in Scotland after the early 1930s, except, of course on the LMS standard classes.

Variations on a single theme These views of CR 60 class No. 14643, GSWR 381 class No. 14663 and the HR 'Castle' class No. 14684 "Duncraig Castle" demonstrate some of the variations in approach, merely in the outside cylinder 4-6-0 field, which the LMS, somehow had to sort out after 1922.

Photomatic

Chapter 1
Caledonian Railway — Introduction and Passenger Tender Classes

Introduction

The Caledonian was undoubtedly the proudest independent railway in Scotland during the pre-1923 period, although followers of the somewhat larger North British Railway (later part of the LNER) might disagree. They were, undoubtedly, rivals for public favour but the Caledonian could claim *primus inter pares* if for no other reason than that it owned the Scottish end of the very first Anglo-Scottish route, the so-called 'West Coast line' between London and Glasgow. Interestingly, from the LMS standpoint, although the West Coast influence on locomotive affairs went into something of a decline during the 1920s in England, the same was not true north of the border where Caledonian precepts remained supreme, much to the disgust of the Glasgow and South Western Railway — *see Chapters 5 & 6*. The third Scottish LMS constituent (the Highland — *see Chapter 7*) was less affected in this respect.

In part, this was due not simply to the numerical superiority of the Caledonian 'fleet' vis-à-vis the other two Scottish systems in the LMS group; but also to the, more or less, uninterrupted continuity of locomotive development at St. Rollox (Caledonian) from 1882 onwards, unmatched by either the G&SWR or the HR.

The earliest of the many companies which eventually formed the Caledonian system (incorporated in 1845) were the Glasgow, Garnkirk and Coatbridge (GGC) and the Dundee and Newtyle lines, both opened in 1831, so it is hardly surprising that the Caledonian, as heir to both these pioneer developments, felt itself on a 'par' as it were, with the LNWR in England which had incorporated the trend-setting Liverpool and Manchester Railway of 1830 in its system. Be that as it may, however, the Caledonian 'proper', as we may term it, was incorporated in 1845 to link Carlisle with the Scottish capital, Edinburgh and, by virtue of a take-over of the GGC *(above)*, with Glasgow also. This objective was achieved in 1848 and, in the same year, the Scottish Central Railway (acquired later by the Caledonian in 1865) reached Perth. In 1866, the Scottish North Eastern became part of the Caledonian and this extended influence to Aberdeen.

Further acquisitions followed and a general rationalisation of activity, but almost from the very start of things, the Caledonian controlled the principal Carlisle to Aberdeen 'axis' with its Glasgow and Edinburgh connections. By the time of the Grouping, the Caledonian had not only purloined the Scottish Royal Coat of Arms for its own use (without permission from the Lyon Court as far as is known), but also, with fine conceit, styled itself 'The True Line', and by then had reached well into the Scottish Highlands with its Callander and Oban line. Consequently, with the formation of the LMS, the 'Caley', as it was familiarly known, felt itself able to 'call the shots', to use a modern phrase. In fact, the CR did not, for reasons of legal technicalities, officially become part of the LMS system until mid-1923, some six months after the official 'Grouping'. In this, a parallel may be drawn with the North Staffordshire Railway *(see Volume Two)*.

In locomotive affairs, although the LMS inherited some Caledonian engines from the pre-1882 period, the vast bulk of the fleet was strongly based on the ideas laid down by Dugald Drummond when he took over at St. Rollox in that year. Although Drummond did not occupy the top position as long as his successors, he was in command for long enough (1882-91) to set the Caledonian firmly on its way. Fortunately, his work on the 'Caley' was, in retrospect, far more significant for posterity than his subsequent efforts for the London and South Western Railway, during which time he tended to lose his way somewhat. Regrettably for the LMS, his younger brother Peter followed Dugald's precepts even after he had 'gone astray', as it were — and this was to have somewhat unfortunate consequences for the Caledonian's arch rival in the LMS camp, the G&SWR, when Peter Drummond took over at Kilmarnock — *see Chapters 5 & 6*.

However, as far as the CR was concerned, Dugald Drummond left at the height of his powers and reputation, and his successor, Lambie, was quite content to develop Drummond ideas during 1891-5. Lambie was, in fact, preceded by the admirable Hugh Smellie (ex-G&SWR — *see Chapters 5 & 6*) who scarcely had time to 'get his feet under the table', as it were, before he died, prematurely. After Lambie, the main locomotive job went, in turn, to John McIntosh, who was very much versed in the Drummond tradition and developed it for almost the next twenty years (1895-1914). Thus, the Drummond-Lambie-McIntosh continuum was dominant in the new-formed LMS, and was not much diluted by the post-1914 developments of the final Caledonian locomotive chief, Pickersgill, who held sway until 1923 — *see Plates 1 to 4*.

After the Grouping, such was the dominance of what one might call the 'Drummond influence' that there was already the basis of a quasi-standardisation in Scotland (for pre-grouping engines) on Caledonian methods, and the LMS took full advantage of this. Several of the Highland classes (e.g. the 'Loch' and 'Ben' class 4-4-0s and the 0-6-0 goods engines) could accept Caledonian pattern replacement boilers so the LMS 'adopted', for want of a better word, most of the St. Rollox 'ground rules'. This was bad news for the G&SWR *(see Chapter 5)* but helped prolong the life of the Highland Railway fleet *(see Chapter 7)*.

In consequence, many Caledonian types survived almost intact throughout the LMS period — and well into BR as well. They were robustly built and mostly well-suited for the duties they performed so, not surprisingly, there were many areas in Scotland where the new LMS 'standard' influence took some time to make itself felt.

Thus it was that, as time went by, the characteristic LMS 'look' in Scotland became increasingly a mix of Caledonian

Plates 1 to 4 These four views have been put together to demonstrate the evolutionary continuity of line displayed by most Caledonian passenger engines. No. 14113 (ex-CR No. 1196) shows a small-wheeled Drummond 4-4-0 of the 80 class, dating from the late 1890s (LMS livery Code A1) while No. 14331 is a McIntosh 'Dunalastair II' of some eight years or so later (LMS livery Code A1). Some eight or nine years after this, many of the same visual characteristics were still present on the McIntosh 903 class, of which the famous 'Cardean' is seen in LMS guise (livery Code B7). Finally, No. 14476 shows a Pickersgill 4-4-0 (ex-CR No. 937) of some ten years later, still carrying most of the Caledonian 'hallmarks' as it were — LMS livery Code B4.

Authors' Collection and G. Coltas

and post-group practice as far as locomotives were concerned, and by nationalisation, this had, more or less, spread well beyond the purely Caledonian lines to embrace virtually the whole of the LMS Northern Division.

As is customary in our pre-group surveys of the larger companies, we have split our detailed survey into the four principal categories identified at the great 1923 renumbering, but before starting this detailed analysis, it is necessary to make a general preliminary comment about Caledonian locomotive styling, particularly in the matter of tenders.

From the outset it should be made clear that one of the reasons for the visually strong 'house' style of most Caledonian engines was the fact that the basic Drummond philosophy was little modified by either Lambie or McIntosh. Moreover, although the two later engineers introduced their own new designs (frequently by nothing more than enlargement of the earlier Drummond types) they also continued to build the basic Drummond designs — particularly in the realm of 0-6-0 goods engines. However, each successive batch of new tender engines was, not infrequently, accompanied by a new design of tender, even though the engines themselves scarcely changed in appearance. Consequently, by the Grouping, there were probably more subtle variations of tender style to be seen on the ex-Caledonian engines than on any other comparably-sized fleet inherited by the LMS.

Now it is not part of our remit to give a full history of Caledonian locomotives — and in any case, much of the data can be extracted from the sources given on *page vii* — but, since our main concern is with visual differences, we think it helpful to analyse the CR tender variations as a preliminary to the individual class summaries. In this context, we would very much like to thank our good friend Duncan Burton of Edinburgh for preparing the summary which we give below. This should prove more than adequate for understanding the LMS situation. En passant, we should point out that the quoted locomotive class designations follow the Caledonian system of identification — i.e. using the CR running number of the first member of the class to enter service.

Caledonian Tenders *(see also Summary Table 1)*
by Duncan Burton

The designs of Dugald Drummond commencing in 1883 formed the basis of all subsequent Caledonian tender development. Most locomotive classes retained their original types of tender but confusion arises where two groups of tender are concerned viz: (1) the Drummond 2,500 gallon tender with underhung springs, and (2) the McIntosh tenders with bogies.

The Drummond tenders (type D1) were originally fitted to the two 4-4-0 passenger classes with 6ft. 6in. and 5ft. 9in. wheels and to the 0-6-0 goods engines, but within a few years new tenders of greater water capacity (types D2 and D3) were designed for the passenger engines and their old tenders went to new goods engines then under construction. There matters rested until the scrapping of the passenger engines which commenced a few years prior to the Grouping when some, at least, of their original tenders were also transferred to the goods engines to replace those of the older underhung spring variety which, it was said, were more prone to hot journals. Later, in the 1930s a number of 3,000 gallon tenders from scrapped McIntosh goods engines (34 class) and also from the 55 Class Oban Bogies and a few of the Pickersgill 300 class engines became available, displacing more of the old tenders on the long-lived Drummond 'Jumbos'. Tender changing within this class probably occurred to a limited extent before the Grouping but became more prevalent thereafter, being further confused when the LMS tender numbering scheme was applied and tank capacity plates were fitted. This was because the building dates on the numberplates and the water capacity sometimes corresponded with the class of tender which the locomotive had had originally, rather than that to which it was now coupled!

The LMS history of the bogie tenders is less complicated and commenced when six wheel tenders of the 3,570 gallon style from scrapped engines became available. The reason for the demise of the bogie tenders is not hard to find apart from the obvious one that maintenance costs would naturally be greater on account of the additional wheel set and complexity of running gear. As the post-grouping period advanced, the locomotives concerned, which had originally been provided with high capacity bogie tenders for long non-stop runs, were progressively relegated to less important work for which these large heavy tenders were no longer necessary. The replacement tenders were nearly all of the 3,570 gallon type M3 from the 0-8-0 mineral engines (600 class) and the mixed traffic and goods 4-6-0s (908, 918 and 179 classes) whose tenders became available from the late 1920s onwards, together with a few older tenders of the same capacity (type M1) from the 721 class 'Dunalastair I' and 13 class (type L1), which survived to be similarly utilised. The much newer tenders of the 3-cylinder 4-6-0s of the 956 class (type P3) withdrawn in the early 1930s, were also saved and used in the same way.

By the mid-1930s most of the bogie tenders had gone as, by that time, with the disappearance of many of the non-superheated engines, there were enough of the second-hand six wheel tenders to go round, and the surviving bogie tenders mainly of the final design (type M10) fitted to the superheated 'Dunalastair IV' class were in due course withdrawn.

Note: The various styles of tender are summarised at *Table 1*, together with one plate reference to each type denoting where they are illustrated in this survey. Further tender data is also available on *page 204* (and elsewhere) of the book *Forty Years of Caledonian Locomotives* by Campbell-Cornwell, which, we think, will adequately cater for most readers' needs. Finally, if necessary, we shall refer to tenders in the text, captions, etc. by the reference letters and numbers given at *Table 1*.

TABLE 1

Summary of Caledonian Railway Tenders

Compiled by D. Burton & D. Jenkinson, March 1985

A : Six wheel type   *Note: Wheelbase of all types, except M4/P4, was 13ft., equally divided; M4/P4 were 11ft. wheelbase*

Designer	Date	Water Capacity (gall.)	Type Reference	Plate Reference	Visible Distinguishing Points	Locomotives to which Coupled	Other Remarks
Drummond	1883	2,500	D1	Plate 171 Chapter 4	Long slots between axleboxes, underhung springs, tool box at rear on footplate	1. 294 class 0-6-0 ('Jumbos') built between 1883-9 2. 66 class 4-4-0 } until replaced 3. 80 class 4-4-0 } by type D2	Rear tool boxes removed by LMS. Tank capacity varied between 2,500 and 2,840 gallons but rationalised at 'nominal' 2,500
Drummond	1888	2,840	D2	Plate 12 Chapter 1	Long slots between axleboxes, springs over axleboxes, no toolbox, tank to rear of footplate, shallow depth frames as D1	1. 66 class 4-4-0 } replacing D1 2. 80 class 4-4-0 } type which then went to 294 class Some D2 type subsequently to goods engines (when 4-4-0s were scrapped), replacing older D1 type	Tank capacity properly 3,130 galls. but for some reason they became known as 2,840 gallon type and were thus classified by LMS
Drummond	1889	3,560	D3	Plates 14/15/17 Chapter 1	Short slots in pairs between axleboxes, springs over axleboxes, long rear overhang (no toolbox), deeper frames than D1/D2	66 class 4-4-0 built 1889-91. After scrapping of 4-4-0s, some used to replace type D1 on 294 class 0-6-0	Tank capacity variously quoted as 3,540/50/60. Later classified as 3,560 gallons
Drummond	1889	2,500	D4	Plate 10 Chapter 1	Long slots between axleboxes, springs over axleboxes, toolbox at rear on footplate, deeper frames than D1/D2 and same depth as D3 type	1. 294 class 0-6-0 2. 80 class 4-4-0, but probably none left with this type at the Grouping	Some of these originally quoted as 2,840 gallons, all later classified as 2,500. Rear toolboxes removed by LMS
Lambie	1893	3,570	L1	Plate 16 Chapter 1	Slots in pairs between axleboxes, frames deeper, rear overhang slightly less than D3 (no toolboxes), tank sides higher than D3	1. 13 class 4-4-0 — Lambie version of 66 class Some tenders to 294 class 0-6-0, also 4-4-0 types, replacing bogie tenders	A short-lived design, not very numerous and quickly evolved into type M1
Lambie	1893	2,800	L2	Plate 13 Chapter 1	Long slots between axleboxes, similar to D4 but with higher tank sides, toolbox at rear	294 class 0-6-0 — a few transferred to 812 class class in BR days	Rear toolboxes removed by LMS

Designer	Date	Water Capacity (gall.)	Type Reference	Plate Reference	Visible Distinguishing Points	Locomotives to which Coupled	Other Remarks
McIntosh	1895	3,570	M1	Plates 18/19 Chapter 1	Slots in pairs between axleboxes, generally as L1 but side plates at front higher to match engine cab	721 class 4-4-0 — 'Dunalastair I'. When engines scrapped, some M tenders replaced bogie types on other 4-4-0s	Difficult to identify from L1 type
McIntosh	1899	3,000	M2	Plate 175 Chapter 4	Slots in pairs between axleboxes, lower tank sides, deeper top coping and wider platform than M1 type	1. 812 class 0-6-0 2. 30 class 0-6-0 3. 34 class 2-6-0 } some later to 294 class 0-6-0, in replacement of old D1	There is evidence that 4-4-0 No. 146 ran for a short time in CR days with this type
McIntosh	1901	3,570	M3	Plate 57 Chapter 1	Slots in pairs between axleboxes, generally as M2 but noticeably higher tank sides	1. 600 class 0-8-0 2. 908 class 4-6-0 3. 918 class 4-6-0 4. 178 class 4-6-0 } when locos scrapped, most replaced bogie types on 4-4-0s. After scrapping of 4-4-0s some then went to 812 class 0-6-0	Transfer of type M3 to 812 class 0-6-0 allowed further type M2 to go to 294 class 0-6-0, by which time (c. 1947/8) tender changing was becoming rather indiscriminate
McIntosh	1902	3,000	M4	Plate 54 Chapter 1	No slots between axleboxes, short wheelbase, coalrails, height as type M3	55 class 4-6-0 — 'Oban Bogie' when locos scrapped, tenders to 294 class replacing more D1 types	
Pickersgill	1916	4,200	P1	Plates 45/74 Chapter 1	Slots in pairs between axleboxes, generally as M3 but higher and wider tank	1. 113 class 4-4-0 2. 60 class 4-6-0 (CR & LMS built)	LMS built tenders were 3,500 gallon, some with water scoops and tank vents
Pickersgill	1918	3,000	P2	Plate 199 Chapter 4	Slots in pairs between axleboxes, generally as M2 but with coalrails, front side plates higher to match engine cab	300 class 0-6-0 — a few early loco withdrawals in late 1930s allowed tenders to go to some 294 class 0-6-0s	During latter days there was some tender changing between 300 and 812 class 0-6-0s
Pickersgill	1921	4,500	P3	Plate 77 Chapter 1	Slots in pairs between axleboxes, similar to P2 but wider with top coping less sharply angled in consequence. Footplate width reduced at front to match engine	956 class 4-6-0 When engines scrapped, tenders to 4-4-0, replacing bogie type	The numerically smallest CR design of tender
Pickersgill	1922	3,000	P4	Plate 68 Chapter 1	No slots between axleboxes, generally as type M4, including coalrails, but wider	191 class 4-6-0 — 'Oban Bogie'	Difficult to identify from M4 type

B : Eight wheel type

Designer	Date	Water Capacity (gall.)	Type Reference	Plate Reference	Visible Distinguishing Points	Locomotives to which Coupled	Other Remarks
McIntosh	1897	4,125	M5	Plate 22 Chapter 1	Inverted springs with equalising beam between each pair of bogie axleboxes — overhang at rear less than at front	766 class 4-4-0 — 'Dunalastair II'. Late survivors had tenders replaced by six wheel type	Two tenders briefly paired (experimentally) with rebuilt 66 class 4-4-0s before the grouping
McIntosh	1899	4,125	M6	Plate 30 Chapter 1	Similar to M5 but nearly equal front/rear overhangs and deeper top coping	900 class 4-4-0 — 'Dunalastair III'. Replaced by six wheel type in LMS period	Somewhat difficult to identify from M5 type
McIntosh	1903	5,000	M7	Plates 60/62 Chapter 1	Similar to M6 but much higher tank sides and wider	49 class 4-6-0 — 'Sir James Thompson' type	
McIntosh	1904	4,300	M8	Plate 38 Chapter 1	Similar to M6 but somewhat higher tank sides — width as type M7	140 class 4-4-0 — 'Dunalastair IV'. Replaced by six wheel type in LMS period	A sort of intermediate type between M6/M7. Difficult to identify from M6
McIntosh	1906	5,000	M9	Plates 61/63 Chapter 1	Similar to M7 — see remarks	903 class 4-6-0 — 'Cardean' type	Visibly identical to M7 but 2 tons heavier
McIntosh	1910	4,600	M10	Plate 39 Chapter 1	Similar size to M7/M9 but with independent springs to each axlebox	139 class 4-4-0 } superheated 39 class 4-4-0 } 'Dunalastair IV' Replaced by six wheel type in LMS period from later 1930s	Last bogie tenders to survive. Division of types M8/M10 between classes did vary at times

Passenger Tender Classes

With a handful of exceptions, the Caledonian passenger tender engines which came to the LMS were all of Drummond or later design but, collectively, they displayed more variety than any of the other three principal LMS defined categories. Apart from the great family of 4-4-0s, stretching through from Drummond to Pickersgill, most CR tender classes were built in small batches, mostly of different classes, which did not show quite the same degree of progressive evolution as did the 4-4-0s. The latter wheel arrangement was by far the most characteristic Caledonian type right to the end, and the somewhat later, and rather tentative move to the 4-6-0 was never quite so successful as, for example, on the LNWR.

As usual, we deal with the subject in ascending LMS number order, except for the 4-6-0 types where we deal with all the McIntosh designs before those of Pickersgill.

4-2-2 No. 14010; Power Class 1, later 1P

This celebrated locomotive, now preserved in the National Collection as CR No. 123 at the Glasgow Transport Museum, was one of a pair of engines built for the Edinburgh International Exhibition of 1886. The other one was 4-4-0 No. 124 (allocated LMS number 14296) — *see below*. Both engines, very much in the Drummond idiom, were built with the backing of the Caledonian Railway. The 4-2-2 was built by Neilson & Co. and the 4-4-0 by Dübs & Co. Both were awarded Gold Medals.

The 4-2-2 was the only 'single' in Caledonian stock in 1923 — and, indeed, after the end of the Connor period (1881), the CR never really perpetuated the single driver type on its own account; yet No. 14010 was destined to become the last 4-2-2 on the LMS, outlasting all of the far more numerous ex-Midland examples *(see Volume 4)* and was not withdrawn until 1935.

During LMS days, the engine was used widely for engineers' saloon workings, somewhat analogous to the ex-LNWR 'Cornwall' — *see Volume Two, Chapter 4*, and was successively painted in both full crimson livery and the 1928 lined black style. In *Plates 5 to 7* is given a selection of pictures during this period.

Plate 5 LMS No. 14010 (ex-CR No. 123) in full crimson livery, Code A1, on engineers' duty in the 1920s.
Authors' Collection

Plates 6 & 7 These two views show No. 14010 at Perth in 1932, carrying lined black livery, believed Code B6 but possibly B3. Note the retention of smokebox numberplate and smokebox wingplates, the decorative 'star' on the smokebox door and the non-standard tender with 'D'-shaped frame slots. Compared with *Plate 5*, the boiler is a replacement.
Authors' Collection and L. Hanson

Brittain Oban Bogie 4-4-0 (LMS Nos. 14100-7; Power Class 1, later 1P)

These, visually, highly distinctive engines were eight survivors of a class of ten such engines built in 1882 to work trains on the Callander and Oban line. They had light axle loading and small (5ft. 2in.) driving wheels, but proved highly successful. Around the turn of the century — roughly halfway through their lives — they were all rebuilt and it was in this form that the LMS received them. They were the only pre-Drummond passenger tender engines handed over by the CR and only four of the eight were actually renumbered (Nos. 14100/3-5). All four received the correct red livery, usually with 14in. figures *(Plates 8 & 9)*, but one of them (No. 14103) received the larger size numerals — *see Volume One, Plate 178*.

The last example was scrapped in 1930 and none are believed to have received the post-1927 livery.

Plates 8 & 9 Opposite side views of Brittain 'Oban Bogies' Nos. 14100 (ex-CR No. 1179) and 14105 (ex-CR No. 1186), both bearing red livery, Code A3.

Authors' Collection and A. G. Ellis

Drummond 80 class 4-4-0 (LMS Nos. 14108-15; Power Class 2, later 2P)
These engines, variously known as the 'Coast', 'Clyde' or 'Gourock' Bogies, were not the first Drummond 4-4-0s, but their small driving wheels (5ft. 9in.) caused them to be placed ahead of the main 6ft. 6in. series in the LMS lists. They were, however, in all essentials, a small-wheeled version of Drummond's pioneering 66 class *(below)*.

The engines were built in two batches of six each in 1888 and 1891 respectively, the second batch originally having boilers designed by Hugh Smellie. Of the eight which reached the LMS, five were from this later series (Nos. 14111-5), the other three being from the original Drummond batch. All eight had been rebuilt in the 1903-8 period, and their essential character is well represented by *Plates 10 & 11*.

Plate 10 Drummond 80 class 4-4-0 No. 1081 (allocated LMS No. 14109 but never carried) remained in Caledonian colours until withdrawn in 1930. It is shown here, somewhat grubby, in 1926, tender type D4.

A. G. Ellis

Plate 11 Full red livery, Code A1, was carried by 'Coast Bogie' No. 14114 (ex-CR No. 1197) when photographed early in LMS days with tender cab and devoid of wingplates, tender type D2.

Authors' Collection

As with the Oban Bogies *(above)* only four were renumbered (Nos. 14110/3-5) and all four received the correct pre-1928 livery with 18in. figures, Code A1. We do not believe any received the 1928 style of painting but one of them (No. 14115) later received 14in. tender side figures, at which time it also carried a tender cab *(Plate 12)*. There was, also, at least, one mismatched engine/tender combination early in the LMS period *(Plate 13)*.

The last survivors were withdrawn in 1930, by which time, one at least, No. 14113, had received a replacement tender type D3 *(Plate 14)*. For other tender details see *Table 1*.

Plate 12 The somewhat rare 14in. numerals (Code A3) were carried by No. 14115 (ex-CR No. 1198) for a while in the 1920s. This view also gives a very good impression of the Drummond 2,840 gallon tender, type D2.

A. G. Ellis

Plates 13 & 14 CR No. 1196 became LMS No. 14113 but is seen in the first view in 1926 with a type L2 black tender from 0-6-0 No. 17399. As No. 14113 it received full red livery, Code A1, and, very unusually for this type, a 3,560 gallon tender, type D3.
A. G. Ellis and Photomatic

Drummond/Lambie 66 class 4-4-0 (LMS Nos. 14290-310; Power Class 2, later 2P)
This series of express engines represented the true 'seed-corn' out of which grew the whole of the succeeding Caledonian 4-4-0 family, including the 80 class *(above)* and, in particular, the 'Dunalastair' family *(below)*. By LMS times, their numbers had been somewhat depleted from the full total of 35 (all but the last six being of Drummond origin) and several variations in appearance and detail were to be observed. The total of 35 also included the Edinburgh Exhibition 4-4-0 No. 124 *Eglinton* (allocated LMS number 14296 but never carried) which was, in effect, assimilated to the 66 class, although slightly different in detail when built by Dübs in 1886. The final six Lambie engines were sometimes differentiated as the 13 class *(see Table 1)* and originally given different tenders. The engines were built in five batches between

Plate 15 Few of the genuine Drummond-built 66 class 4-4-0s in original condition received new LMS numbers. No. 1083 should have become LMS No. 14305 but was withdrawn in 1928 in the condition shown here. The tender is type D3.
Stephen Collection, courtesy NRM

Plate 16 No. 14310 (ex-CR No. 1018) was a genuine Lambie example, and is seen here in full red livery, Code A1, paired with its original tender, type L1.

A. G. Ellis

1884 and 1894, plus No. 124 itself, and, to some extent, the second and successive batches displayed some variations in detail on the previous series if only in matters of tenders, so they are not the easiest of engines to sort out *(Plate 15)*. However, by LMS days, the survivors numbered 18 from the Drummond period (Nos. 14290-307) and three from Lambie's time (Nos. 14308-10). Of these last three, one of them (No. 14309) — *see title page* — was a sort of composite engine, extensively rebuilt in 1907, embodying bits and pieces from both Drummond and Lambie engines. Having rather more Lambie vintage parts in it than Drummond, it was officially regarded as one of the later series! Visually, the Lambie engines *(Plate 16)* were very like the Drummonds, save for their fire box mounted safety valves.

By 1923, a further visual complication was evident. Originally, all the engines had displayed the characteristic Drummond 'lines' as exemplified by *Plate 15*, but during the McIntosh period, ten of them had been rebuilt with his larger 'Dunalastair I' type boiler and new cab and splashers as in *Plate 17*. In this form they were visually very like the 'Dunalastair III' series *(below)*, and in work terms they were all but indistinguishable from the 'proper' 'Dunalastairs', so much so that they were often nicknamed 'Dunalastair Drummonds'. The survivors of these ten rebuilds were allocated LMS numbers 14298-304 and, just to make confusion complete, the first rebuild (allocated LMS number 14300 but never carried) retained its original Drummond pattern leading splasher, incorporating the prominent front sandbox — so it looked more like a 'Dunalastair I' than a McIntosh rebuild!

Plate 17 LMS No. 14302 (ex-CR No. 1075) displays the 'Dunalastair-Drummond' styling with McIntosh cab/splashers. Livery Code is A1 and tender type, D3.

Authors' Collection

Already nearly 40 years old at the Grouping, the class as a whole did not last long, and only a few received new LMS numbers viz:

>Original Drummond type: 14290/7
>Rebuilt 'Dunalastair Drummond': 14298/301-2/4
>Original Lambie type: 14308-10

The last survivor was No. 14304, withdrawn in 1931.

In livery terms, we have confirmed all but Nos. 14290 and 14301 in full crimson lake, Code A1, and none at all in the post-1928 style. We see no reason to believe that Nos. 14290 and 14301 were not finished in the red livery, and it may be that a few late survivors were repainted lined black. If so, Nos. 14298/304/9-10 would be the possible contenders.

Finally, the one CR named example, No. 1079 *Carbrook*, which was repainted by the LMS as No. 14297, lost its name in the process. Six wheel tenders were attached throughout but there were some changes *(see Table 1)*.

McIntosh 'Dunalastair' series 4-4-0 (LMS Nos. 14311-65, 14430-60; Power Class 2/3, later 2P/3P)
These celebrated engines — possibly amongst the most famous in the land — are one of the most difficult groups of engines to disentangle at this range in time. So, craving the indulgence of those who know their 'Caley', an attempt is made to offer a simplified summary before discussing their LMS history.

There were four series of engines, each being a progressive enlargement of the preceding type and the first of the series (the 'Dunalastair I') was itself a development of the Drummond/Lambie 66 class. At a later stage, after the introduction of superheating, all but the 'Dunalastair I' locomotives had a superheated version as well, generally by rebuilding. However, the fourth series (the most numerous), included 21 examples built new with superheaters. In summary, therefore, the situation inherited by the LMS was as follows:

Type	*Saturated engines (Class 2/2P)*	*Superheated engines (Class 3/3P)*
'Dunalastair I'	14311-25	None
'Dunalastair II'	14326-36	14430-3 (rebuilds)
'Dunalastair III'	14337-48	14434-7 (rebuilds)
'Dunalastair IV'	14349-65*	14438-9 (rebuilds)
		plus: 14440-60 (built new)

NB: one of these (No. 14356) was also superheated — *see page 24*

The name 'Dunalastair' arose from that given to the first of them all (CR No. 721, LMS No. 14311) and when the second series was built, CR No. 766 (LMS No. 14430) was, for a time, named 'Dunalastair II'. The third and fourth series never carried any nomenclature but popular usage caused it to be applied. Only a few were named at all in CR days, and the LMS removed the one remaining name on repainting in 1925 — as was the case with all ex-CR named engines. For the record, the full list of names was:

>CR No. 721 *Dunalastair** — LMS No. 14311 ⎤
>CR No. 723 *Victoria** — LMS No. 14313 ⎬ all 'Dunalastair I' type
>CR No. 724 *Jubilee** — LMS No. 14314 ⎦
>CR No. 766 *Dunalastair II** — LMS No. 14430 Dunalastair II (superheated) type
>CR No. 779 *Breadalbane* — LMS No. 14335 Dunalastair II type
>
>* *Names removed by CR during 1914-7*

From this point onwards, we shall deal with the whole group in their four individual series.

a) 721 class — 'Dunalastair I' type (LMS Nos. 14311-25)
These engines *(Plates 18 & 19)*, were built in 1896 as a direct enlargement of the 66 class with which they shared many visual features in common (front splashers/sandbox, cab style, etc.). The LMS did not, materially, change their appearance except for the removal of any residual smokebox wingplates and occasionally fitting replacement 'pop' safety-valves. In *Plates 18 to 21*, a good cross-section of their post-grouping characteristics is represented. No. 14316 *(Plate 21)* was the only example with a tender cab.

All fifteen were renumbered and we believe all were painted in the full red livery in the style of that shown in *Plate 18*. We have confirmed ten of them *(below)*. Scrapping started in 1930 and the last two went in 1935, so we do not believe that all received the later LMS livery. We give a few confirmed examples below but feel sure there were more.

The engines ran throughout with McIntosh six wheel tenders (type M1 — *Table 1*), apart from the odd possible short term changes, and some of these tenders survived the engines to replace tenders on other classes.

Plates 18 & 19 These similar views show 'Dunalastair I' locomotives Nos. 14319 (ex-CR No. 729) and 14312 (ex-CR No. 722) in full crimson and early lined black liveries respectively, Codes A1 and B5. Note the retention of front numberplate after the livery change. The tender type is M1 in both cases.

Authors' Collection

Plate 20 'Dunalastair I' No. 14318 (ex-CR No. 728) is seen in close-up wearing very grimy lined black livery, Code B7.

Photomatic

Plate 21 No. 14316 (ex-CR No. 726) was the only 'Dunalastair I' with a tender cab and is carrying lined black livery, Code B7.

Authors' Collection

Livery Samples

Code A1	14311-2/4-7/9-23
Code B5	14312
Code B7	14313/5-6/8

} all with 53in. letter centre spacings

b) 766 class — 'Dunalastair II' type (LMS Nos. 14326-36; 14430-3)

The 'Dunalastair II' locomotives dated from 1897-8 and were a straightforward enlargement with bigger boilers and smokeboxes, extended frames and a wider cab. Visually, they presented quite a difference in style from the 'Dunalastair I' type, although self-evidently still 'in the family'. The prominent leading splasher/sandbox of the earlier 4-4-0s gave way to a more graceful and conceivably more 'modern-looking' arrangement, while the newer cab was altogether more substantial. Add to this the well-proportioned and new style McIntosh bogie tender (type M5 — *Table 1*) and the whole ensemble presented a beautiful visual balance, well represented by *Plates 22 & 23, also page iv*.

Plate 22 No. 14328 (ex-CR No. 770) typifies the 'Dunalastair II' series as originally reaching the LMS. The cabside cut-away displays the earlier McIntosh shape, characteristic of the 'Dunalastair II' locomotives. The livery Code is A1 and the bogie tender is type M5.

Authors' Collection

Plate 23 The handsome lines of the unmodified 'Dunalastair II' are well displayed in this view of No. 14327 (ex-CR No. 768), livery Code B7, in the early 1930s, still with its bogie tender.

Authors' Collection

Fifteen were built, four of which were superheated between 1914 and 1918. These four could be identified mostly by the somewhat more aggressive front end with its slightly longer smokebox and modified front frame shape, and cylinder covers *(Plate 24)*. Other than this basic difference, the engines themselves changed little during early LMS days.

Plate 24 Superheated 'Dunalastair II' No. 14431 (ex-CR No. 769) shows off the front-end characteristics of the superheated series and still carried its bogie tender in the 1935/6 period when this view was taken — livery Code B7.

Photomatic

Plate 25 No. 14326 (ex-CR No. 767) was one of several 'Dunalastair II' locomotives to be modified by the LMS — *see main text*. It is seen carrying post-1927 livery Code B6. The black shaded transfers are quite clear on the original picture. It is paired to a replacement six wheel tender, type M1.

Authors' Collection

Rather later in the LMS period, some further rebuilding took place. Between 1930 and 1933, Nos. 14326/9/32/6 were given higher pitched 812 class 0-6-0 boilers with Pickersgill chimneys and domes. The cab sides were raised at the top, and this coincided with the change from bogie to six wheel tenders *(below)*. A somewhat similar rebuild was carried out on No. 14433 of the superheated series, this time using a Pickersgill boiler from a 300 class 0-6-0. An LMS period rebuild is shown in *Plate 25*.

The tender changing alluded to above affected the whole group of engines and we try to indicate this in our summaries. Quite soon after the Grouping, two at least (Nos. 14326 and 36) ran for a while with the larger final type of bogie tender *(Plate 26)*, while during the 1930s, all seem to have been given six wheel tenders, mostly of the Lambie/McIntosh 3,570 gallon series (types L1/M1/M3 — *Table 1*). The likelihood is that some came from withdrawn 'Dunalastair I' engines and the altered engine/tender appearance consequent upon these changes is quite well shown in *Plate 25*.

Plate 26 A large type bogie tender, type M10, originally paired with the superheated 'Dunalastair IV' series is seen attached to saturated 'Dunalastair II' No. 14336 (ex-CR No. 780), circa 1926 — livery Code A1 — but see also *page iv* for a view of No. 14336 with original bogie tender.

Authors' Collection

Plates 27 & 28 Nos. 14430 (ex-CR No. 766) and 14431 (ex-CR No. 769), livery Codes A1 and B2 respectively, were two superheated 'Dunalastair II' locomotives which carried their smokebox wingplates well into the LMS period.

Authors' Collection and A. G. Ellis

In livery terms, all engines are thought to have received red liveries and we have confirmed all but two — *overleaf*. During the 'red' phase, the bogie tenders were universal and many engines received the lined black style before the tender changes took place. The characteristic smokebox wingplates seem mostly to have gone from the saturated engines before LMS days, but some of the superheater rebuilds kept them for quite a while *(Plates 27 & 28)*. We believe that all four had wingplates at the Grouping and have only failed to confirm No. 14432 thus fitted in early LMS days. All were removed later.

During the lined black period, 12in. or 14in. figures seem to have been the common choice with possibly more 12in. examples to be seen. One or two engines seem to have been given 10in. figures early after the change in style, and the 14in. figures appear to have gone somewhat out of fashion when the six wheel tenders began to appear. A handful of survivors (all from the saturated series) ran on into the war years, but we do not believe many became plain black before scrapping, although some did seem to receive the typically Northern Division 10in. figures in the late 1930s.

Livery Samples

	Saturated Series	*Superheated Series*
Code A1	14326/14326§/8-34/6/6§	14430-3
Code B2	14334	14431★
Code B3	14326-8/32*/5-6	14433★
Code B4	14335★	14433★
Code B6	14326*/9*/34/6*	14433
Code B7	14326★/7★	14431★
Code B8 (C21)	14333	

Notes:
 * * LMS period rebuilds — see text
 * § Large bogie tender fitted — pre-1928
 * ★ Retaining original bogie tender — post-1927
 1. Letter spacing 53in. between centres — post-1927
 2. Although the change was gradual, this summary assumes six wheel tenders post-1927 (unless indicated) and standard bogie tenders pre-1928

c) 900 class — 'Dunalastair III' type (LMS Nos. 14337-48; 14434-7)

The sixteen 'Dunalastair III engines dated from 1899-1900 and although somewhat bigger than the 'Dunalastair II' type, they were, visually, somewhat difficult to distinguish from the second series. The main changes were an increased coupled wheelbase from 9ft. to 9ft. 6in., a larger firebox and a higher pitched boiler. The latter caused the chimney to be marginally shorter in height. The cabside cutaway embodied a continuous sweeping curve of rather more graceful appearance than the compound curves of the cabside of the 'Dunalastair II' type, but this was a somewhat subtle visual distinction. The bogie tenders were also slightly different, although of the same tank capacity (type M6 — *Table 1*).

The superheated rebuilds dating from 1914-18 were modified in a like manner, relative to the originals, as had been the earlier engines, and in *Plates 29 & 30* we give the two versions, more or less 'as received' by the LMS. By this time, the saturated engines had received Pickersgill boilers of the same size, fitted with 'pop' safety-valves. Wingplates had mostly gone before 1923 with this series and the rest soon after. When tender changing began, the six wheel replacements were of the same sort as given also to the 'Dunalastair II' type.

Plates 29 (Left) & 30 (Top Right) These two views, photographed from much the same angle, show the saturated and superheated versions of the 'Dunalastair III'. No. 14342 (ex-CR No. 892) is in red, Code A1 and superheated No. 14436 (ex-CR No. 901) is lined black, Code B4. The tender type is M6 in both cases. Note the continuous curve of the cab cutaway — one of the several minor changes from the 'Dunalastair II' series.
Authors' Collection

All told, therefore, there was a great deal of visual similarity between the second and third series of McIntosh 4-4-0s, and this extended to their liveries in LMS days. As far as we know, all were originally repainted red and most are confirmed *(overleaf)*. After the livery change, lined black was universally applied with the 12in. numerals predominating but not exclusive. The few confirmed examples with 14in. figures seem mostly to date from the early 1930s before the tender changing, but there was at least one exception *(Plate 31)*. The 12in. style *(Plate 32)* embraced both tender periods, and one or two late survivors achieved the characteristic plain black with 10in. figures *(Plate 33)*.

Plate 31 No. 14341 (ex-CR No. 891) received its replacement six wheel tender, type M3, while still bearing 14in. figures — Code B4, possibly B7.

Authors' Collection

Plate 32 'Dunalastair III' No. 14348 (ex-CR No. 899) in 1932 at Dalry Road with 12in. figures and bogie tender — livery Code B6.
L. Hanson

Apart from No. 14435, destroyed in the Dinwoodie smash of 1928, withdrawal took place between 1937 and 1948. Interestingly, No. 14435 (superheated series) was the pioneer member of the class (CR No. 900).

Livery Samples

	Saturated Series	*Superheated Series*
Code A1	14338/41-7	14434-7
Code B3	14341/5/7	14436/7★
Code B4	14341 (possibly B7)	14436★/7★
Code B6	14339/42/8★	—
Code B7	14343★	—
Code C21	14340	14434

Notes:
★ *Retaining original bogie tender — post 1927*
1. *Letter spacing 53in. between centres — post-1927*
2. *Assume six wheel tenders (post-1927) and original bogie tenders (pre-1928) except where indicated. The changeover period was quite drawn-out.*

Plate 33 The last 'Dunalastair III' was superheated No. 14434 (ex-CR No. 894). It is seen in wartime livery Code C21 and fitted with a type M1 tender plus a snowplough for working over the ex-HR lines.
Authors' Collection

d) 140 class — 'Dunalastair IV' type (LMS Nos. 14349-65; 14438-60)

The fourth and final 'Dunalastair IV' design totalled 40 examples, almost as many as the first three added together. They were even more imposing-looking, once again embodying larger (and higher-pitched) boilers, extended wheelbase, and even more impressive bogie tenders. These tenders were either type M8 or the even bigger type M10 *(see Table 1)*. The latter generally ran with the superheated engines and the type M8 with the saturated series — but there were exceptions. As usual, tenders were exchanged in the 1930s and the picture coverage embraces this change-over.

Visually, the main distinguishing points of the 'Dunalastair IV' engines (other than their absolute size) were the re-shaped cab spectacle glasses, the continuous single coupling rod splashers (slightly depressed in the centre), and the noticeably less tall boiler mountings. There were the usual front end differences between the saturated and superheated types and both versions are shown in Plates 34 & 35, much as they came to the LMS.

Plates 34 & 35 These two views show the saturated and superheated 'Dunalastair IV' types with the two different styles of bogie tender, types M8 and M10 respectively. No. 14365 (ex-CR No. 136) was the last of the saturated engines and No. 14445 (ex-CR No. 117) was built new with a superheater. Both are in LMS red livery, Code A1.
Authors' Collection

Plate 36 No. 14356 (ex-CR No. 147) was a superheated 'Dunalastair IV' in the saturated number series. Note that although the smokebox/cylinder covers reflect the superheated style, the side frames still betray their 'saturated' origin. The livery code is B3.

L. Hanson

The 'Dunalastair IV' type dated from 1904 but, by the time the final series was built in 1910, and later, superheating was standard, and slightly more than half were delivered in superheated condition from new — *see summary on page 14*. All told there were, originally, 19 saturated and 22 superheated examples. One of the latter (CR No. 121) was destroyed in the Quintinshill disaster of 1915. A few of the saturated examples were subsequently superheated. Two of these went into the proper part of the LMS number list (Nos. 14438-9), but one other which had also received a superheater was, for some reason, put in the 143XX series. This was No. 14356 *(Plate 36)*.

Liveries were much the same as for the earlier series and details need not be repeated, except to remark that since the bulk of the superheated series survived well beyond LMS days, most of them eventually became plain black during the final LMS period, while many acquired BR painting styles as well. The pictures shown in *Plates 37 to 44* encompass much of the story in this respect and the summary *(page 27)* gives a pretty good LMS period cross-section. The last survivor was No. 54458 (ex-CR No. 41) which lasted until the end of 1957, more than twenty years after the first withdrawals (saturated series) in early 1937. The earliest withdrawals were over thirty years old when they went, and the last ones were nigh on fifty. They were good engines.

Plate 37 Superheated 'Dunalastair IV' No. 14441 (ex-CR No. 132) in LMS red, Code A1, still with smokebox wingplates and tender type M10 at Perth, circa 1926.

Authors' Collection

Plate 38 Superheated No. 14438 (ex-CR No. 923) was one of the original saturated series and retained its original pattern type M8 tender in early LMS days — livery Code A1.

A. G. Ellis

Plate 39 By contrast with No. 14438, saturated No. 14358 (ex-CR No. 149) was paired with the later type M10 tender in the mid-1930s — livery Code B3.

Authors' Collection

Plate 40 This view shows the normal engine/tender pairing of the superheated 'Dunalastair IV' in the early 1930s. No. 14448 (ex-CR No. 120) displays 14in. figures, believed to be red shaded, Code B4.

Authors' Collection

Plates 41 to 43 In later LMS days, the 'Dunalastair IV' locomotives received replacement six wheel tenders and began to display various combinations of visibly rivetted smoke-box, new chimney and new dome cover. The 10in. numerals also began to appear, and these three pictures attempt to cover this period. Saturated No. 14363 (ex-CR No. 137) has a new chimney/smokebox and type M1 tender — livery Code C21, superheated No. 14447 (ex-CR No. 119) has a new dome cover, smokebox and type M3 tender, but retains its original chimney and has lined black livery — Code B8, while superheated No. 14452 (ex-CR No. 45) shows all four changes simultaneously — livery Code C21.

Authors' Collection

Plate 44 This early BR view shows No. 54438 (ex-CR No. 923) in plain black with only its new number to betray nationalisation. Interestingly, the smokebox numberplates returned in BR days. The picture makes an interesting comparison with the earlier view of the same engine seen in *Plate 38*.

Authors' Collection

Livery Samples

	Saturated Series	*Superheated Series*
Code A1	14349/51/3-4/8-9/61-2/4★/5	14356§/438/§/9§/41-2/5/7-53/5/7-8/60
Code B3	14349§/53§/7/8★/64	14356§/439§/40★/1★/2★/5/7★/9§/52★/3§/5★/6★/9★
Code B4	—	14440★/14440/3★/6★/8★/51/60★
Code B5	—	14449★
Code B6	14350★/63	14460§
Code B7	—	14442★/6★/50★/3★/6★/9★
Code B8	—	14443/7/54
Code C21	14363	14439-41/3/4★/7/9-55/7-9

Notes:
★   *Bogie tender type M10*
§   *Bogie tender type M8*
1. *Letter spacing 53in. between centres — post-1927*
2. *Assume bogie tender type M8 (saturated series) and type M10 (superheated series) for all pre-1928 references, except where indicated*
3. *Assume six wheel tenders post-1927 except where indicated. The changeover period was quite drawn-out*

Pickersgill 113 class 4-4-0 (LMS Nos. 14461-508; Power Class 3, later 3P)
If proof was needed that the Caledonian Railway never really forsook the 4-4-0 type, then the 48 Pickersgill engines of the 113 class should serve the purpose. More numerous, even, than the 'Dunalastair IV' type, they might even have been regarded as 'Dunalastair V' — so much did they draw on previous development.

They were, in essence, a modified version with larger cylinders of the superheated 'Dunalastair IV', and visually very similar. The only really obvious changes were the front frame shape above the footplate, the straight-topped coupling rod splasher and the large six wheel tender (type P1 — *Table 1*). The latter was still obviously in the same line of evolution that had started in the Drummond days and much of the rest was in the McIntosh idiom — unlike some of Pickersgill's other designs. So, right to the end of the line, the 'Caley' 4-4-0 displayed its 'line of breeding', as it were, *(Plates 45 & 46).*

Plates 45 & 46 These two fine views show Pickersgill 4-4-0s Nos. 14473 (ex-CR No. 934) and 14493 (ex-CR No. 88) resplendent in the early LMS livery, Code A1. Note the late retention of old type safety-valves on No. 14473. The tenders are of course the Pickersgill standard, type P1.
Authors' Collection

The engines were built between 1916 and 1922. All came to the LMS and all survived more than ten years in BR ownership before planned scrapping commenced in 1959. This was a rapid process, in the event, and all had gone before the end of 1962. There was one premature accident victim in 1953 after the Gollanfield collision (No. 54481), but apart from this solitary example, every member of the class gave well over forty years of revenue service. Along with the superheated 'Dunalastair IV' type, they well-embodied the quasi-standardisation on CR practice which typified the Northern Division of the LMS *(Plates 47 to 51)*.

Plate 47 A few early Scottish area repaints to the 1928 style came out with small cabside numerals and 'reversed' power class insignia as shown here on No. 14494 (ex-CR No. 89). Interestingly, the power class was marked P4 and not 3P! Insignia were hand-painted and we have coded the style B5. Some sources (unconfirmed) reckon that a few of these early repaints were red, not black, but we can only be certain of a few ex-Highland examples — *see Chapter 7*.
Authors' Collection

Plates 48 & 49 Pickersgill 4-4-0s had a classic cleanness of line, well-shown by these two views illustrating the two most common forms of insignia during the lined black period. No. 14476 (ex-CR No. 937) has unshaded 14in. figures, Code B7, while No. 14496 (ex-CR No. 91) carries 12in. red countershaded figures, Code B3, possibly the most common single variant.
Authors' Collection

Plate 50 The retention of numberplate and smokebox 'star' were unusual features of No. 14489 (ex-CR No. 84) in a somewhat grimy state — Code B3.

Authors' Collection

Plate 51 The somewhat uncommon lined black livery with 10in. figures (Code B8) is shown on No. 14480 (ex-CR No. 75). Most engines carrying this style of insignia were plain black.

Authors' Collection

Plates 52 & 53 Two BR variant painting styles on No. 54499 (ex-CR No. 68) and No. 54506 (ex-CR No. 95) reflect the longevity of the Pickersgill 4-4-0s. Note the chimney, dome and smokebox alterations. No. 54506 is typical of many of these engines in the mid-1950s.
Authors' Collection

The LMS hardly changed them at all in visual terms, and they mostly retained the original tender type throughout. In later years, some of them received rivetted smokeboxes and/or replacement chimneys and dome covers *(Plates 52 & 53)* but they generally retained a very consistent appearance to the end of their lives. The picture coverage and livery summary *(overleaf)* does, we feel, give a good representative cross-section of the type.

Livery Samples

Code A1	14461/3/5-6/8/70-1/3-4/7/85/88-95/7/503-5/7
Code B3	14464/6-70/3/82/6/9/95-7/9-500/3-4/7-8
Code B4	14463/9-71/4/6-7/9-81/501
Code B5★	14494/8/505
Code B7	14461/72/83/7/91/504-5
Code B8	14471/80/91
Code B9	14508
Code C21	14461-3/70/4-5/7-81/4-7/9-91/4-6/8/502-3/5/7

Notes: ★ *Power class markings wrongly numbered and reversed viz: 'P4' not '3P' — an early post-1927 feature*
Letter spacing 53in. between centres — post-1927

Caledonian 4-6-0 Types

The Caledonian Railway introduced many numerically small series of 4-6-0 engines but, always, one had the feeling that its efforts in this direction were somewhat tentative. Even the most celebrated of them — the 903 class or *Cardean type* — were never all that much, if any, better than the four-coupled locomotives for express work.

The smaller-wheeled freight and mixed traffic series were possibly more useful than the purely express designs, but were never particularly numerous. Not surprisingly, therefore, the 4-6-0s, although generally newer than most of the 4-4-0s, did not last quite as well in LMS days. Ironically, the best 4-6-0 class which the Caledonian possessed was probably the series of engines purchased from the Highland Railway — the so-called 'Rivers'.

In order to aid understanding, we have, in this section, elected to deal with the engines by designer rather than in strict LMS numerical order. This is largely because the 4-6-0 was the one area where Pickersgill departed most noticeably from the traditional Caledonian style.

McIntosh 55 class 4-6-0 (LMS Nos. 14600-8; Power Class 3, later 3P)

McIntosh's first essay into the 4-6-0 arrangement emerged in 1902 because of the particular circumstances of working the Oban line. Five of the class came out that year and another four in 1905. They were, in effect, a lengthened version of the 'Dunalastair III' but embodying only 5ft. driving wheels. They were also given distinctive six wheel tenders without frame slots between the axleboxes (type M4 — *Table 1*).

With their small wheels, they proved true mixed traffic engines and normally worked between Stirling, Callander and Oban. As usual, Pickersgill began to remove the smokebox wingplates circa 1917, and none are thought to have reached the LMS still carrying this feature *(Plates 54 & 55)*. Apart from the gradual fitting of 'pop' safety-valves, nothing further of real visual significance was changed until 1930, when the LMS rebuilt two of them (Nos. 14606 and 14607) with larger boilers and cabs, previously used on the 918 class *(see Chapter 4)*; although the cabside-sheets had to be shortened by 6in. below the cut-away. On one of them, the tender sideframes were given small frame openings *(Plate 56)*. In rebuilt form they were very similar to the 918 class.

Plate 54 This lovely picture shows 55 class 4-6-0 No. 14603 (ex-CR No. 58) in full red livery, Code A1. It also shows off the type M4 tender, purpose-built for the class. Note the coal rails and lack of central frame apertures.

Stephen Collection, courtesy NRM

Plate 55 This somewhat inferior picture shows the opposite side of the 55 class pioneer No. 14600 (ex-CR No. 55) in lined black with 14in. numerals, Code B7. Note the replacement 'pop' safety-valves and the somewhat closer than normal spacing (for ex-CR tender engines) of the letters 'LMS'.
Authors' Collection

Plate 56 Reboilered 55 class No. 14606 (ex-CR No. 52) in grubby, lined black livery, Code B3. Note the modified tender side frames — unique to this engine.
Authors' Collection

In livery terms the LMS always regarded them as purely passenger engines, 5ft. wheels notwithstanding, and we believe all became red. We have confirmed six of them. After 1927, lined black was the choice with a slight preponderance of 12in. numerals rather than 14in. However, as usual, we feel that it may have been a case of 14in. early in the lined black period followed by 12in. figures later.

The engines were scrapped between 1934 and 1937, the rebuilt No. 14606 being the last survivor.

Livery Samples

Code A1   14600/2-4/7-8
Code B3   14600/4-5/6§/7§ } letter spacing about 48in. between centres.
Code B7   14600/3-4/7-8

§ After rebuilding with larger boiler — see main text

McIntosh 908 class 4-6-0 (LMS Nos. 14609-18; Power Class 3, later 3P)
In 1906, McIntosh built two series of what might be styled 'intermediate size' 4-6-0s embodying a shorter version of the *Cardean* boiler (903 class — *below*) and smaller wheels. One series (the 918 class) had 5ft. wheels and the LMS classed them as goods engines *(Chapter 4)* while the 908 class had 5ft. 9in. driving wheels, were regarded as 'mixed traffic', and were put in the LMS passenger lists. Confusingly, the superheated version of the 5ft. 9in. type (the 179 class) was regarded as a goods engine and is, therefore, also covered in *Chapter 4*.

All ten of the 908 class came to the LMS, by which time the two former Caledonian named examples (CR No. 909 *Sir James King* and CR No. 911 *Barochan*) had lost their names. The LMS made no real changes, except for fitting 'pop' safety-valves, and all are thought to have lost their smokebox wingplates before the Grouping. *Plate 57* is typical of the 'as received' condition, and *Plate 58* shows the later form. Tenders were always of the 3,570 gallon variety (type M3 — *Table 1*).

Plate 57 No. 14612 (ex-CR No. 911) in red livery, Code A1, shows the 908 class, virtually as built, except for the removal of wingplates. It also gives an exceptionally clear elevation of the McIntosh 3,570 gallon tender, type M3, fitted to these engines.
A. G. Ellis

Plate 58 'Pop' safety-valves have replaced the earlier types on 908 class 4-6-0 No. 14613 (ex-CR No. 912). This picture is one of relatively few we have seen of ex-CR passenger engines clearly carrying countershaded 14in. numerals with lined black livery. The smokebox door carries considerable evidence of overheating!
Authors' Collection

Plate 59 The new cab on No. 14618 (ex-CR No. 917) clearly shows how the visual character of this member of the 908 class was changed. It was the only one so treated, livery Code A1.
Authors' Collection

There was one 'odd man out', LMS No. 14618 (ex-CR No. 917) — *Plate 59*. In 1910 this engine had received a new cab with side windows in an experiment to give greater crew protection. It remained unique in this respect as far as the 908 class was concerned, but the new style cab was adopted on the superheated 179 class version *(Chapter 4)*.

We believe that the LMS painted all of them red during the 1920s, but can only confirm six *(below)*. Scrapping started in 1930, so some may never have received the lined black. Those that we have confirmed in the later livery all sported 14in. figures, mostly of the black shaded variety; this tends to support our belief that the 12in. figures did not appear regularly on many ex-CR types until the mid-1930s, by which time the 908 class was extinct. The only one to survive beyond 1933 was the unique No. 14618, and it had gone by the end of 1935.

Livery Samples

Code A1	14610-2/4/6/8§
Code B4	14613
Code B7	14609-10/2/5/8§

} 53in. letter centres

§ *Side window cab*

McIntosh 49 and 903 class 4-6-0 (LMS Nos. 14750-5; Power Class 4, later 4P)

The generally effective way in which the 55 class 4-6-0s had gone into service on the Oban line *(above)* prompted McIntosh to build two seemingly gigantic 4-6-0s in 1903 for the increasingly heavy West Coast expresses. These were the 49 class of which No. 50 was named *Sir James Thompson* (LMS Nos. 14750-1). They were the most powerful express engines in Britain at the time, and were too long for the turntables at both ends of the Carlisle to Glasgow main line. Probably for this reason, is was not until 1906 that further big 4-6-0s were built, to a slightly modified arrangement — the 903 class (LMS Nos. 14752-5). These had slightly larger boilers but a smaller cylinder diameter, thus reducing the nominal tractive effort from 24,990lb. to 22,667lb. Five were built to this style, of which the first was the legendary No. 903 *Cardean*. One (CR No. 907) was written off after Quintinshill in 1915, so the LMS only got four of them.

During 1911-12, all seven engines were superheated and received identical sized cylinders, thus becoming virtually alike, except for the few inches in boiler diameter. Both types had massive bogie tenders of which the 903 class version was marginally the heavier (types M7 and M9 — *Table 1*).

During Caledonian days, the whole group was allocated to the heaviest express duties, generally with individual crews, and became a very famous series of machines. They were, like most McIntosh engines, handsome of line and beautifully maintained; but it has to be said that they were, in service, no real improvement on the superheated 'Dunalastair IV' type, and it is not too surprising that their LMS life was somewhat foreshortened. Few significant changes were made, except for the replacement 'pop' safety-valves and the removal of smokebox wingplates from the 49 class. However, and one wonders if it was just for sentimental reasons, *Cardean* itself, although now nameless, was withdrawn as No. 14752, still carrying wingplates in 1930, and was the last of her type to survive. The earlier '903s' are also believed to have carried wingplates until withdrawal.

The 49 class outlived the 903 series by some three years by virtue of using spare parts from the withdrawn 903 class engines.

Plates 60 & 61 The two versions of the McIntosh 6ft. 6in. 4-6-0 are shown in red livery, Code A1. No. 14750 (ex-CR No. 49) was the pioneer example and No. 14755 (ex-CR No. 906), the fourth member of the 903 class, still retains its wingplates. Note the variation between safety-valve casings (both original). The smaller boiler diameter of No. 14750 is barely perceptible and best reveals itself by the (slightly) taller chimney and dome cover.

A. G. Ellis

The six engines of the original seven, received by the LMS, were painted red as shown in *Plates 60 & 61*, while the only two we have confirmed in post-1927 livery are the pioneer engines from each series Nos. 14750 and 14752 *(Plates 62 & 63)*. It is possible that No. 14751 was similarly treated, but we feel certain that Nos. 14753-5 were scrapped carrying the red livery.

Plates 62 & 63 No. 14750 again and No. 14752 (ex-CR No. 903 *Cardean* itself), show the final LMS appearance of these most stately of all ex-Caledonian engines. Note the change to 'pop' safety-valves, the retained wingplates on No. 14752, and the slightly taller dome and chimney of the earlier type. The livery code is B7 in both cases.
Authors' Collection and BR (LMR)

Highland Railway Design 'River' class 4-6-0 (LMS Nos. 14756-61; Power Class 4, later 4P)

In 1915, shortly after Pickersgill had taken over from McIntosh at St. Rollox, the Highland Railway's CME, Mr F. G. Smith, was highly embarrassed when his civil engineer refused to accept a new design of 4-6-0, already built. It became quite a 'cause célèbre' in railway history but the upshot was that the Caledonian Railway agreed to purchase the engines 'off the shelf', as it were, and thus, by accident, acquired one of the most outstanding Scottish 4-6-0 designs — certainly better than anything the CR had designed for itself so far, or even afterwards as it was to transpire.

There were only six of them, and the Caledonian numbered them 938-43; so technically they were the 938 class. However, the Highland Railway had intended to name them after rivers (two, indeed were so delivered in HR colours), so 'River' class they became.

Their outside valve gear, Belpaire fireboxes, and other details made them quite unlike anything the Caledonian had, hitherto, employed, but they were fine machines and Pickersgill, realising that the CR had made an astute purchase, tended to fall under their influence with his 191 and 956 class engines *(below)*. The fact that his did not quite match up to the original was no fault of the 'Rivers' themselves.

Plate 64 The massive nature of the 'River' class 4-6-0 is well portrayed by No. 14761 (ex-CR No. 943), the last of the series and painted LMS red, Code A1.

NRM Collection

Plate 65 An opposite side view of No. 14759 (ex-CR No. 941) shows an early post-1927 lined black livery with 10in figures and 'reversed' power class marking — Code B5.

Authors' Collection

During LMS days, the engines changed very little, and their essential character is well portrayed in *Plates 64 to 67*. All were originally painted red, and thereafter lined black. As usual, numeral height varied after 1927. There were at least two given small figures with 'reversed' power class *(Plate 65)* and a few received 14in. figures before 12in. became more common. The tenders were, of course, specific to this type but not altogether dissimilar to the final Pickersgill type. The LMS withdrew the engines only slowly (between 1936 and 1946) and, interestingly, eventually put them to work on the Highland lines for which they had been built in the first place.

Plates 66 & 67 These two close-ups show 'River' class 4-6-0s No. 14759 again — this time in livery Code B3 and a cab-end view of No. 14760 (ex-CR No. 942) in a grubby state, with 14in. figures on the cabside, Code B4. The shading of the insignia on both these pictures is hard to detect, except under a glass and, not for the first time in these volumes, draws attention to the difficulty of resolving the issue in all cases.

BR (LMR)
and Authors' Collection

Livery Samples

Code A1	14756-61 (the whole class)	
Code B3	14756-9	
Code B4	14758-60	53in. letter centre spacing
Code B5	14758-8	
Code B7	14760-1	

Pickersgill 191 class 4-6-0 (LMS Nos. 14619-26; Power Class 3, later 3P)

Although Pickersgill was content in many ways to continue the traditional St. Rollox line of evolution in the 0-6-0 and 4-4-0 field, when it came to 4-6-0s, things were a bit different. In part, this may have been because the McIntosh 4-6-0s were not entirely up to expectations, but we are inclined to think that, in this context, Pickersgill was not uninfluenced by the nature of the excellent 'River' class engines purchased from the Highland in 1915 *(above)*. Certainly, his 191 class engines had many visual features reminiscent of the 'Rivers', but on a smaller scale, and they were, in our view, most attractive-looking engines. Eight of them were built in 1922 just before the Grouping, so they really spent most of their life as LMS engines. They were built for the Oban line, partly to improve on the 55 class *(above)*, but also to replace the ageing Brittain 4-4-0s. In the event, some of them went to Balornock for use on Glasgow area locals, and they did not prove very popular, most of the men preferring the older 55 class engines for the Oban line. Like the latter engines, the 191 class also had their own unique tenders (type P4 — *Table 1*).

The LMS painted them all red — and most attractive they must have looked — *Plates 68 & 69*. After the livery change, one or two received small 10in. cabside figures with the power class marking reversed *(Plate 70)*; thereafter the universal lined black with 12in. figures gradually became dominant *(Plate 71)*. As usual, some of the earlier lined black repaints received 14in. figures (mostly black shaded) in the early 1930s.

Plates 68 & 69 These two excellent pictures show opposite side views of members of the 191 class in full red livery, Code A1. The picture of No. 14619 (ex-CR No. 191), the pioneer example, also illustrates clearly the 'short' Pickersgill tender, type P4, built for these engines, while No. 14622 (ex-CR No. 194) cannot have been long out of shops when this view was photographed. The relatively small boiler size is readily apparent.
A. G. Ellis and Authors' Collection

Plate 70 No. 14624 (ex-CR No. 196) illustrates particularly well the early 'Scottish' version of the 1928 lined black livery, Code B5, with 'reversed' power class markings, P4 rather than 3P! Note the somewhat closer than usual spacing of the 'LMS' and that, being centred on the tank side, the 'M' is well to the rear of the centre axlebox — cf. No. 14600 (Plate 55).

A. G. Ellis

Plate 71 191 class, No. 14626 (ex-CR No. 198) was the last of the series and carries the quite common 12in. counter-shaded figures, Code B3, in the later 1930s. The shading is only visible under magnification. The huge pile of coal on the tender probably indicates either preparation for a long run, or less than perfect thermodynamic capability!

BR (LMR)

Scrapping took place between 1939 and 1945, but we have no records of any of them ever getting the small 10in. figures which began to reappear in Scotland circa 1938. We feel sure that one or two must have been so treated (either with or without lining); perhaps readers can help?

Livery Samples

Code A1	14619-26 (the whole class)	
Code B3	14620-3/5-6	
Code B4	14619	letter spacing about 48in. between centres
Code B5	14624	
Code B7	14623-4/6	

Pickersgill 60 class 4-6-0 (LMS Nos. 14630-55; Power Class 4, later 4P)

If the Pickersgill 191 class *(above)* and 956 class *(below)* had more than a hint of the 'River' class in their visual lines, then his third 4-6-0 design (actually the first to appear, in 1916) had distinct McIntosh visual overtones, outside cylinders notwithstanding. This is hardly surprising since McIntosh himself had schemed out an outside-cylindered version of the *Cardean* type before he retired. In the event, it did not appear (and neither did his proposed 4-6-2), but in 1914, Pickersgill resurrected the notion, changing the specification slightly, to incorporate 6ft. 1in. driving wheels. The 60 class was the result — a neatly-proportioned, almost handsome design still recognisably in the 'Caley' tradition.

Authorised in 1915, the engines did not appear until 1916 — no doubt because of wartime difficulties — and only six were built by the CR itself (LMS Nos. 14650-5). Later writers have given them an indifferent reputation, but they cannot have been as awful as some folk would have us believe. They were no speed merchants but their maintenance was, apparently, of a quite economic order, largely due to their robust construction and large journals. At all events, the LMS saw fit to authorise further production — a somewhat rare privilege for a pre-grouping design if it was not Midland inspired! In consequence, a further twenty appeared during 1925-6 and, for some reason, they took numbers ahead of the CR batch (14630-49), unlike the post-1923 0-4-4Ts *(Chapter 2)* which were numbered after the Caledonian and GSWR series. The tenders of the LMS built series were slightly modified — *see Table 1*

The 60 class were, for most of their life, known as 'Greybacks' for no well-authenticated reason, as far as we can determine. There is a somewhat scurrilous explanation in Glasgow and South Western circles that 'greyback' was a colloquial word for a louse, but we leave readers to their own views!

The fact that there were, all told, 26 of these engines made them the most numerous pre-group Scottish-designed 4-6-0 type, and almost all survived to the BR period. There were only three LMS period withdrawals (Nos. 14632/3/55) and fourteen received BR numbers (Nos. 54630/4-6/8-40/2/7-51/4). Coupled at all times to their original tenders of type P1 *(Table 1)*, their appearance hardly changed, except for a gradual move to visibly rivetted smokeboxes — e.g. *Plate 75*. Consequently, during LMS times, livery was the main variable.

It is known that all the LMS series were outshopped in the pre-1928 red style, and we believe all six ex-CR examples were likewise treated *(Plate 72)*. Our summary, however, only gives confirmed examples. After the livery change, lined black was adopted and again, all available evidence suggests that 14in. figures were used soon after 1927 (usually black-shaded) followed by the widespread adoption of 12in. figures (gold and usually countershaded) which gave way to the 10in. yellow figures, shaded red in the late 1930s and during the war years when plain black livery was adopted. Our picture coverage *(Plates 73 to 76)* attempts to illustrate all significant variables.

Plate 72 The last of the CR-built 60 class 4-6-0s was No. 14655 (ex-CR No. 65) seen here in near original condition at Perth, circa 1926. Note the safety-valve cover. The livery code is A1.
Authors' Collection

Plates 73 & 74 Apart from their running numbers, the LMS-built examples of the 60 class could also be identified by the 3,500 gallon tender, often with prominent tank vents on the tender rear. This is well-shown on Nos. 14645 and 14631, and the latter view also shows well the characteristics of the LMS built tender. Both engines are lined black but No. 14631 has countershaded insignia (Code B4) whereas No. 14645 has black shading. Once again, one needs a magnifying glass and the original prints to establish these facts!

Authors' Collection and W. L. Good

Plates 75 & 76 Visibly rivetted smokeboxes eventually became universal on the 60 class as did smaller than 14in. numerals. No. 14644 has 12in. black shaded figures and lined black livery, Code B6, while No. 14639 is in plain wartime black with 10in. yellow figures, red shaded, Code C21. Interestingly, No. 14644 still carries its front numberplate even though the Carlisle (12A) shedplate is post-1935.

*Photomatic
and A. G. Ellis*

Livery Samples

Code A1	14630-2/4/6-8/47-9/50-1/3-5
Code B3	14630/3/41/3/7/50/4-5
Code B4	14631/45/8/51/4
Code B6	14638/44/9
Code B7	14632/5-7/41-2/7
Code C21	14632/9/40/2/5/52

} letter centre spacing about 53in.

Plate 77 No. 14803 (ex-CR No. 959) was the highest number allocated to an LMS passenger tender engine at the Grouping and, it cannot be denied, was displayed on an imposing machine, the last of the 956 class engines. Comparison with *Plate 64* reveals the considerable similarity to the 'River' class, including the high-sided 4,500 gallon tender, type P3. The livery code is A1.

A. G. Ellis

Pickersgill 956 class 4-6-0 (LMS Nos. 14800-3; Power Class 5, later 5P)
These four intriguing locomotives appeared in 1921 as the largest passenger engines ever built for the 'Caley'. They had three cylinders and derived inside valve gear — an almost unheard of combination outside Gresley circles — but, by all accounts, were a less than perfect design. The derived motion was incredibly complex (to circumvent patents perhaps) and quite a bit of detail was based on the 'River' class, which they somewhat resembled, albeit with round-topped fireboxes *(Plate 77)*.

A study of reports of the trials with No. 956 in 1921 suggests that the engines were not as bad as some writers have maintained; but they were disappointing, and the fact that there were only four of them would hold little appeal to the new-style LMS management. Prior to the Grouping, all but No. 956 had been fitted with independent Stephenson valve gear for the middle cylinder, retaining the original outside Walschaerts linkages for the other two cylinders.

The LMS simply did not want to know about them. They were, presumably, allowed to run only until the boilers fell due for replacement and/or major overhaul, and were withdrawn during 1931-4. All four were painted red and all received identical versions of the post-1927 lined black with 14in. black shaded numerals, Code B7. Both versions are shown in *Plates 77 & 78*. Their 4,500 gallon tenders (type P3 — *Table 1*) were unique, and went in turn to 4-4-0s when the engines were scrapped.

Plate 78 We conclude our survey of Caledonian passenger tender engines with this well-known official view of No. 956 herself, as LMS No. 14800 in lined black livery Code B7. Its massive proportions were not displeasing, even if the performance capability was not all that was wanted.

BR (LMR)

Chapter 2
Caledonian Railway — Passenger Tank Classes

Just as it placed great faith in 4-4-0 passenger tender engines, so too did the Caledonian concentrate on one particular wheel arrangement in the passenger tank series — the 0-4-4T. All but 26 engines were of this arrangement, but divided into several categories. In the survey which follows, we shall divide this series into its principal sub groups for ease of understanding. As usual, we tackle the engines in LMS renumbering order.

Drummond 262 class 0-4-2ST (LMS Nos. 15000-1; Power Class later 0P)
In 1885, Drummond built two 0-4-2STs for working the Killin branch of the Callendar and Oban line. They were, in effect, elongated versions of the 264 class 0-4-0ST *(see Chapter 3)* with enhanced coal capacity, and we cannot quite understand why they were not put into the freight lists. The LMS certainly always painted them plain black.

Known as the 'Killin Pugs' both received LMS numbers, although some sources state that No. 15000, scrapped in 1928, was never renumbered. We have seen a picture of it in livery Code C4 (panel on bunker, numbers on cab) but too poor to reproduce. No. 15001 also received a pre-1928 livery (Code C4) this time with the red panel on the tank and number on the bunker. Again, the picture is too poor for reproduction.

No. 15001 lasted until 1947 and was, after 1927, painted as shown at *Plates 79 & 80*, carrying Code C13 and Code C21 in succession.

Plates 79 & 80 These two views of the longer-lasting of the Killin Pugs No. 15001 (ex-CR No. 1263) show livery Codes C13 (with Ramsbottom safety-valves) and C21 (with 'pop' safety-valves). Note also the replacement chimney in the later (April 1946) view.
Authors' Collection and H. C. Casserley

Plate 81 Lambie 4-4-0T No. 15025 (ex-CR No. 6) in red livery, Code A1, at Glasgow, circa 1926.
Authors' Collection

Lambie 1 class 4-4-0T (LMS Nos. 15020-31; Power Class 1, later 1P)
These extremely neat and attractive engines were the only type peculiar to the Lambie period, owing nothing to a Drummond predecessor. All but the last two (LMS Nos. 15030-1) were renumbered. Built in 1893, they were little changed when the LMS got them 30 years later, except that their original condensing apparatus had been removed from 1917 onwards.

The LMS painted most of them red at first *(Plate 81)*, and we can confirm six so finished, viz: Nos. 15020-2/4-5/9. There may have been one or two more, but CR No. 7 (later LMS No. 15026) remained blue until after the livery change in 1928, going straight from CR colours to LMS black livery. Another example, LMS No. 15027, received the freight livery, Code C2.

After the livery change, these engines received their own unique version of the 'St. Rollox' livery — lined black with 18in. tank side figures. The bunker was too small to position the 14in. 'LMS', so the circular emblem was retained. This made the layout of insignia precisely the same as in the red period, and it is often difficult on a black and white picture to tell them apart — *Plate 82*. However, we have every reason to believe that Nos. 15023-6 were finished in this way.

Plate 82 The St. Rollox lined black livery style, Code B1, was carried by 4-4-0T No. 15023 (ex-CR No. 4). Note the identical insignia layout to the red example in *Plate 81*.
A. G. Ellis

Plate 83 No. 15028 (ex-CR No. 9), clearly wearing 14in. countershaded numbers applied in the St. Rollox style in 1931. Other details are obscure, but we feel the engine must have been black with red lining. We have not coded this variation.

Authors' Collection

In 1931, No. 15028 received a different version of this livery, employing 14in. countershaded numerals — *Plate 83*. We showed it with 18in. numerals in *Plate 185 of Volume 1*, but whether it was red or black at that time is impossible to judge. We have no record of any of these engines carrying a recognisably conventional version of the proper 1928 livery, and the last one (No. 15025) was withdrawn in 1938.

Drummond 171 class 0-4-4T (LMS No. 15100-14; Power Class 1, later 1P)
These engines, introduced in 1884, were the oldest ex-CR 0-4-4Ts at the Grouping and very few actually received LMS numbers (Nos. 15103-4/7-8/10/4). With their dome-mounted safety-valves and distinctive solid bogie wheels, they were, visually, full of character *(Plate 84)*. As received by the LMS, all were carrying redesigned boilers dating from 1908. The later examples had larger bunkers (LMS Nos. 15106-14) and, in 1924, two of the early ones (LMS Nos. 15103/4) were rebuilt with new boilers; No. 15104 also receiving a larger bunker at the same time *(Plate 85)*.

We only have livery records of three of them, Nos. 15103-4/14. All received red livery, Code A3, and we illustrate the two known post-1927 liveries in *Plates 85 & 86*. Most were withdrawn in the 1920s, but the 1924 rebuilds lasted until 1933 (No. 15104) and 1944 (No. 15103).

Plate 84 Red-painted Drummond 0-4-4T No. 15103 (ex-CR No. 1177), livery Code A3, shortly after rebuilding in 1924.

A. G. Ellis

Plate 85 This crisp view of 0-4-4T No. 15104 (ex-CR No. 1178) clearly shows the lined out black livery, Code B7. Examination of the original clearly indicates the use of black shaded transfers and the presence of white 'highlight' lines on the numerals.

BR (LMR)

Plate 86 No. 15103 was the last survivor of the Drummond 0-4-4T series. It is seen here with replacement 'pop' safety-valves and a visibly rivetted smokebox. The transfers are gold with black shading, and there is just a trace of lining visible on the bunker side under a glass — livery Code B6.

Photomatic

Lambie 19 class 0-4-4T (LMS Nos. 15115-24; Power Class 2, later 2P)

These engines, sometimes attributed to McIntosh, were the first of the, soon to become very numerous, 5ft. 9in. 0-4-4Ts. They actually appeared in 1895, almost co-incident with the Lambie/McIntosh transition, and embodied the same boiler as the 4-4-0T *(above)*, and the 0-4-4 wheel arrangement was presumably developed from the Drummond type. They were given condensers for working the new 'low level' suburban lines in Glasgow.

When the LMS received them they had been recylindered, had coal rails on the bunker and the condensing apparatus had gone (from 1917 in the latter case); but they were still substantially as built *(Plate 87)*. During 1932-5, the LMS rebuilt a number of them with McIntosh boilers of the later 0-4-4T type. These were Nos. 15115/8-22 but with no significant visual change.

Plate 87 This grubby view of No. 15118 (ex-CR No. 22) shows a Lambie 19 class 0-4-4T in the St. Rollox livery — Code B1. The lack of clarity of the paint scheme does not prevent the picture revealing the neat lines of this first series of 5ft. 9in. 0-4-4Ts.
W. L. Good

All were renumbered but we have been quite unable to find evidence of any of them carrying the red livery. We feel certain that some must have done and would welcome more details. After the livery change, the St. Rollox style was applied to some of them, thereafter the conventional 1928 lined black. There was no real consistency in numeral size until plain black (Code C21 — *Plate 89*) became widespread. Scrapping commenced in 1946 and all but three reached BR, five of them being renumbered (55119/21-4). As far as we can judge, they were regarded in LMS/BR days no differently from the genuine McIntosh series, and lasted just as well *(Plate 90)*.

Plate 88 19 class 0-4-4T, No. 15116 (ex-CR No. 20) in lined black livery, Code B7. Note the flatter-topped dome cover at the time (1937—Dumbarton).
Authors' Collection

Plate 89 The wartime livery, Code C21, and a buckled front end on No. 15117 (ex-CR No. 21). This engine was withdrawn in 1948 without renumbering.
A. G. Ellis

Plate 90 No. 55119 (ex-CR No. 23) shows a 19 class 0-4-4T in early BR livery at Grangemouth in 1952 with a stovepipe chimney, but retaining round-topped dome casing.

Authors' Collection

Livery Samples

Code B1	15118/20★/23
Code B5	15124
Code B6	15119
Code B7	15116★
Code C21	15117/22★/3★

} LMS spacing at about 30in. centres

★ *replacement 'flatter' topped dome cover — see Plate 88*

McIntosh 92 class 0-4-4T (LMS Nos. 15125-46; Power Class 2, later 2P)

The first pure McIntosh 0-4-4T was developed from the Lambie 19 class in 1897, with the same boiler but larger bunker and side tanks. The last ten, built in 1900 (LMS Nos. 15137-46) were sometimes regarded as a separate type (879 class). All had condensers and this apparatus began to be removed from 1917 onwards. The second series could, however, always be recognised by the single front footstep rather than the double footstep of the first series — *compare Plates 91 & 92*.

Detail changes during early LMS days mostly took the form of removing the remaining condensing apparatus, a certain amount of boiler changing — including at least one Drummond example *(Plate 91)* — and the gradual fitting of replacement 'pop' safety-valves, but the older type of valves lasted quite a long time *(Plate 93)*. Later in the LMS period, one or two received a flatter topped dome cover — quite a common practice on ex-CR engines — and right at the end, a somewhat unpleasant looking stovepipe chimney put in an appearance *(Plate 94)*.

LMS liveries were typical for the ex-Caledonian tank engine classes. The red period was followed by lined black, in which instance, quite a number received the distinctive St. Rollox variation *(Plates 95 & 96)*. With the conventional post-1927 lined black livery, 14in. figures were almost always employed, there being plenty of room for them on the bunker side. The 'LMS' letter spacing was always closed-up on these engines to about 30in. between centres, as indeed it was on by far the bulk of all ex-Caledonian tank engines, passenger or freight. All but three of them reached BR, most of them also being renumbered. By this time, most, if not all, were wearing wartime livery — Code C21.

Plates 91 & 92 Nos. 15128 (ex-CR No. 95) and 15138 (ex-CR No. 880) represent the two versions of the 92 class 0-4-4T, both livery Code A1. Note the change in front footstep pattern. Both show interesting features. No. 15128 has received a replacement Drummond boiler with dome-mounted safety-valves, and the bunker side is only partially lagged — normally a pre-1897 practice of Stroudley origin, perpetuated by Drummond. No. 15138 was one of the last to keep condensing apparatus, revealed by the large pipe linking smokebox to tank.
A. G. Ellis and Authors' Collection

Plate 93 No. 15134 (ex-CR No. 101) kept its old type of safety-valves until the 1930s. It is seen in conventional post-1927 lined black (Code B4) — the insignia are undoubtedly gold, countershaded red.
Real Photographs

Plate 94 No. 15130 (ex-CR No. 97) was one of several ex-Caledonian engines transferred to the West Riding after World War II. It carried wartime livery, Code C21, and exhibited both 'flat' dome and stovepipe chimney when photographed near Keighley on a typical LMS stopping train.

BR (LMR)

Plates 95 & 96 These two views show the characteristic St. Rollox livery, Code B1, on members of the 92 class in 1930 (No. 15143, ex-CR No. 885) and 1933 (No. 15126, ex-CR No. 93) respectively. Note the change in dome and safety-valves on No. 15126 — the older of the two engines and with the double front steps.

Authors' Collection and A. G. Ellis

Livery Samples

Code A1	15125/8+/32-3/8-9/44-5
Code B1	15126×/7×/9/43/5
Code B4*	15131/4/7/40-1/5
Code B7*	15125×/30/2/42/4
Code C21	15130×§/1/3/5§/6/40

} letter spacing about 30in. centres

Notes: + Drummond boiler
× 'Flat' topped dome
§ Stovepipe chimney
* There is always difficulty in separately identifying these two types from photographic sources. We give a 'best' estimate where we are in doubt

McIntosh 104 class 0-4-4T (LMS Nos. 15147-58; Power Class 1, later 1P)

These engines, a 4ft. 6in. driving wheel version of the Lambie/McIntosh 5ft. 9in. type, were often known as 'Balerno' or 'Cathcart Circle' tanks, their smaller wheels being designed for quick acceleration between the close-spaced stations on these two routes. They have a strong place in our affection because, back in the early 1960s, they were the root cause of our whole investigation into LMS liveries! One of us had given to a model of a McIntosh 5ft. 9in. engine, the number of a 'Balerno' tank, and was never allowed to forget it. The published information at that time was so 'thin' that we resolved to tackle the problem and it eventually led to this total compilation. However, we digress.

The 'Balerno' tanks were built in 1899, all reached the LMS, and few changes were made to them. They did not last as well as the 5ft. 9in. engines, largely because bridge strengthening and track alterations during LMS days allowed the more numerous 5ft. 9in. engines to operate over the hitherto special preserves of the 104 class. The latter therefore became less useful, they were only twelve in number, and thus became 'non-standard', even by Scottish criteria. The last withdrawal was in 1938 (No. 15153), nine years after the first had gone.

In decorative terms, the usual sequence was followed except that many of the red ones had 14in. figures (Code A3) rather than 18in. figures of the 5ft. 9in. engines *(Plate 97)*. There were, however, many exceptions — *see summary* — and those which received the St. Rollox style of lined black after 1927 *(Plate 98)* always did so with 18in. figures.

Most examples served out their time in conventional post-1927 lined black, generally with 14in. figures *(Plate 99)*, but occasionally with the 12in. variant *(Plate 100)*.

Plate 97 The changed character which the 4ft. 6in. wheels and full width cab/bunker gave the so-called 'Balerno' tanks is well-shown in this view of No. 15150 (ex-CR No. 107) at Dalry Road in red livery, Code A3.

A. G. Ellis

Plate 98 The St. Rollox style livery, Code B1, on 104 class 0-4-4T No. 15153 (ex-CR No. 110). Like many engines photographed at this time (circa 1932), the red lining failed to register on a somewhat scruffy engine, but we have no reason to suppose it was not there.
Authors' Collection

Plates 99 & 100 In these two views, 'Balerno' tanks Nos. 15157 (ex-CR No. 169) and 15150 again, both fail to reveal any lining on the post-1927 black livery, applied conventionally. However, we believe they were given red lining — Codes B7, B6 respectively.
A. G. Ellis and Authors' Collection

Livery Samples

Code A1	15148/51/5/8
Code A3	15147/50-1/4
Code B1	15153/5
Code B6	15150
Code B7*	15149/52-3/7-8

Letter spacing about 30in. centres (applies to Codes B1, B6, B7*)

* One or two of these may have been Code B4 but we cannot be precise

McIntosh 439 class 0-4-4T (LMS Nos. 15159-240/15260-9; Power Class 2, later 2P)

The 439 class was a direct development of the final series of the 92 class engines but without the condensing gear. The boiler pressure was 10p.s.i. higher and tank capacity was slightly raised, but on many railways they would have been put in the same class as their predecessors. Visually they were, for all practical purposes, indistinguishable. They were so successful that they were continued in production both by Pickersgill and by the LMS. The Pickersgill/LMS-built series differed in having a slightly lengthened bogie and coupled wheelbase, but a shorter distance between the two sets of wheels. The total wheelbase remained the same. Four of the Pickersgill engines had even higher boiler pressure and strengthened front buffer planks for banking duty at Beattock. Thus, the total progression was as follows, there being probably more differences here than between the 439 class and its predecessor type:

LMS Nos. 15159-226	Original McIntosh engines — *Plate 101*
LMS Nos. 15227-36	Pickersgill engines built during 1915/22 with altered wheelbase — *Plate 102*
LMS Nos. 15237-40	Pickersgill engines built in 1922 with higher boiler pressure and strengthened buffer planks — *Plate 103*
LMS Nos. 15260-9	LMS-built, 1925, as Nos. 15227-36 — *Plate 104*

We can offer no logical explanation for the gap in number sequence between the end of the ex-GSWR sequence (No. 15254 — *page 134*) and the start of the LMS-built Pickersgill 0-4-4Ts.

Plate 101 (caption overleaf)

Plates 101 (Previous Page) & 102 to 104
The four sub-categories of the 439 class are represented here viz:

> No. 15214 (ex-CR No. 164), McIntosh — livery Code A1
> No. 15235 (ex-CR No. 435), Pickersgill — livery Code B1
> No. 15238 (ex-CR No. 432), Pickersgill banking type — livery Code A1
> No. 15266, LMS-built — livery Code C21

Note the prominent sandbox fillers below the tanks of the Pickersgill/LMS series — a characteristic-distinguishing feature of the LMS built engines, later applied to some of the earlier ones as well.

A. G. Ellis

During LMS days, livery apart, the main changes were the gradual (and usual) introduction of 'pop' safety-valves on the McIntosh engines, some replacement of dome covers with a flatter topped version and, in later years, replacement stovepipe chimneys on some examples. All but two (LMS Nos. 15163/205) reached BR and only four (LMS Nos. 15184/8/90/2) failed to be renumbered plus, possibly, No. 15174. A great number lasted until the early 1960s and one of them (CR No. 419, LMS No. 15189) is, happily, preserved. There could hardly be a more appropriate Caledonian type to represent its suburban activities.

During LMS days, liveries followed the usual trend. The red style was applied universally with 18in. figures as far as we know *(Plate 105)*, but at least one of them had the numerals 'closed up' to avoid the tank side step *(Plate 106)*. There may have been more like this and, being such a large group, a few undoubtedly failed to be repainted until the black livery came in.

Plates 105 & 106 The red livery 'sat' very well on many of the Scottish engines — not least the 0-4-4Ts. Nos. 15168 (ex-CR No. 448) and 15194 (ex-CR No. 127) are seen here in red, Code A1, No. 15194 with close-spaced numerals. These were not very typical but at least missed the tank side step — *see Plate 101*.

A. G. Ellis and Authors' Collection

The change to lined black saw the odd experiment with small 10in. figures *(Plate 107)*; thereafter either the St. Rollox style *(Plate 108)* or 14in. figures in the conventional form *(Plates 109 to 113)* was normal. Most of them eventually became plain black with 10in. figures, yellow with red shading, as for example, No. 15130 of the earlier series *(Plate 94)*. Our summaries give a good cross-section, and we also include a few additional views of some of the more interesting variations — e.g. *Plates 109 & 114*.

Plate 107 This view shows an early lined black repaint (note the retained front numberplate) on 0-4-4T No. 15183 (ex-CR No. 464). The numerals are 10in. and possibly hand-painted — Code B5.
A. G. Ellis

Plate 108 The St. Rollox style, Code B1, when clean, was a distinctly attractive variant of the lined black livery, well displayed on No. 15216 (ex-CR No. 458). To some extent, this treatment existed alongside the conventional arrangement as in *Plates 109 to 113* but, we believe, did not actually make an appearance until just after the first of the orthodox post-1927 repaints.
Authors' Collection

Plate 109 No. 15267 of the LMS-built series was another fairly early 439 class repaint to the post-1927 style, this time with 14in. figures. However, the self-evident tonal change between boiler/smokebox, and the possible indication of lining on the footsteps, strongly suggests that it retained red livery with insignia displayed in Code A7 mode. Unfortunately, the picture, taken at Polmadie in 1932, lacks definition, so we cannot be more positive.

L. Hanson

Plates 110 & 111 These two early black repaints, both with smokebox numberplates, probably date from circa 1929 and display black shaded insignia, Code B7. No. 15170 (ex-CR No. 450) has a conventional layout but, for some reason, No. 15239 (ex-CR No. 433), one of the Pickersgill 'banking' series, has the 'LMS' set too low on the tank sides.

Authors' Collection and A. G. Ellis

Plates 112 & 113 Regular reference is made in this survey to the problem of resolving insignia shading, and these two views help to make the point. No. 15160 (ex-CR No. 440) clearly carries counter-shaded insignia, Code B4, but then so too does No. 15261 of the LMS-built series. The latter, however, only reveals its true state under a magnifying glass. Note that, interestingly, No. 15261 when seen at Edinburgh in 1933, had acquired a boiler still carrying Ramsbottom type safety-valves, no doubt as a result of boiler exchanging at shopping.

BR (LMR) and Authors' Collection

Plate 114 This last picture of the 439 class 0-4-4Ts shows Pickersgill 'banker' No. 15237 (ex-CR No. 431) with all three of the principal detail changes (chimney, dome and 'pop' safety-valves) and painted in plain black wartime livery with 10in. figures. Interestingly, these are clearly black shaded — very unusual — and probably plain yellow (Code C24). It is possible that they were originally red-shaded, and painted over in wartime black at a 'local' area refurbishment.

A. G. Ellis

Livery Samples

Code A1	15168/72/6/94/205/8-9/14/6/22/4/27-9/34/6/8/60-9
Code B1	15159/63/6/70-1×/9×/82/4-5×/7/9/90×/1/4/6/200/2/8-9/11×/2×/5-6/21/31×/5/9/66×/8-9
Code B4*	15160/8-9/97/201/7/13/5/7-20/2-3/4×/6/9/61-2/7+/9
Code B5	15183
Code B7*	15162×/4/7/70/4-5/7/81/232-3/9/60×
Code B8	15159
Code C21	15159×/65×/74×/5-6/82/4/8/91×/2×/3×§/8§/9×/203§/12-3/23×§/30×/4×/5×/40/61/5-6/8
Code C24	15237

Notes: × 'Flat' topped dome
+ Possibly red Code A7 — see Plate 109
§ Stovepipe chimney
* There is the usual difficulty of resolving these two styles — we give a best estimate
Letter spacing, post-1927, at about 30in. centres

Pickersgill 944 class 4-6-2T (LMS Nos. 15350-61; Power Class 4, later 4P)

The one and only 'big' passenger tank designed for the Caledonian was, like that of its West Coast partner, the LNWR, of the 4-6-2 arrangement. The large sized passenger tank was not really a particularly significant pre-group contribution to the LMS fleet, but the Pickersgills were as good a try as anyone achieved until the LMS itself produced its successful 2-6-4 types from the late 1920s onwards. They certainly outlasted the much more numerous LNWR 4-6-2Ts *(Volume 2)*, and were undoubtedly more effective than most of the assorted 4-6-4Ts which the LMS acquired, even if they were not particularly outstanding in purely objective terms.

The Caledonian engines, often referred to as the 'Wemyss Bay' tanks because of their initial utilisation, were introduced in 1917 as a handsome tank engine equivalent of the 60 class 4-6-0s. They did not share precise dimensional equivalence in either wheel diameter, cylinder or boiler size, but the family resemblance was clear to see and, had they been built as tender engines, they would have been revealed as a somewhat smaller version of the 60 class with 5ft. 9in. wheels.

Sadly, they showed less versatility than the 0-4-4Ts once the LMS 2-6-4Ts had become established, becoming increasingly used for humdrum workings and banking duties, and rather less widespread on the passenger services for which they were designed.

In visual terms they scarcely changed at all from their late Caledonian/early LMS appearance *(Plate 115)*. We believe all became red, after which the St. Rollox variation, along with the use of 14in. numerals were the common lined black styles *(Plates 116 to 119)*. We presume that most eventually became plain black but have not confirmed many. Two failed to reach BR (Nos. 15357/8) but, of the rest, only two did not receive 5XXXX numbers (LMS Nos. 15351/5). One or two examples received replacement dome covers of flatter-topped nature *(Plate 120)*.

Plate 115 Pickersgill 4-6-2T No. 15357 (ex-CR No. 951) in red livery, Code A1. Note the open cover to the sandbox filler above the side footsteps.

A. G. Ellis

Plate 116 The St. Rollox style livery, Code B1, on No. 15354 (ex-CR No. 948) — again the sandbox filler cover is open. The picture was taken at Glasgow (St. Enoch) in 1935, ex-G&SWR territory, be it noted.

V. Forster

Plate 117 This view of No. 15361 (ex-CR No. 955) at Polmadie, in the early 1930s, displays the handsome lines of the Pickersgill 4-6-2Ts to good effect — livery Code B7.

Authors' Collection

Plates 118 & 119 These pictures show left-hand side views of the Pickersgill 4-6-2Ts and should help model makers. No. 15360 (ex-CR No. 954) is in a grimy livery — probably Code B7, but No. 15357 clearly carried countershaded characters, Code B4, when at Polmadie in mid-1937.

A. G. Ellis and Authors' Collection

Plate 120A We cannot date this picture (unless one of our readers happens to be the featured schoolboy) but believe it to be wartime, or shorty afterwards. The insignia are likely to be plain yellow, Code C24, much less common than the red shaded variety at this time. Note the flatter top to the dome cover.

Author's Collection

Livery Samples

Code A1	15350-1/3/5/7-9/61
Code B1	15351/4/7-8×/61
Code B4	15357×/9
Code B7	15351-3/5/60-1
Code C21	15352
Code C24	15353x

letter spacing about 53in. centres

x *Flatter topped dome cover*

Plate 120B We take leave of Caledonian passenger engines with this characteristic shot of the pioneer member of the 4-6-2T series, CR No. 944 itself, running as LMS No. 15350 shortly after the Grouping — livery code A1 — at the head of an appropriate type of train. Post-grouping practice is already creeping in — note the ex-MR six wheeler at the back of the receding train.

A. G. Ellis

Chapter 3
Caledonian Railway — Freight Tank Classes

The Caledonian freight tank group consisted of some 250 engines, dominated, numerically, by the 0-6-0 side tanks dating from the very start of the McIntosh period. However, before these began to appear in 1895, the Caledonian had been very much wedded to the saddle tank configuration, and quite a number of these older types survived to LMS days. We therefore propose to depart slightly from the strict LMS numerical order in this section of the book and deal with all the saddle tanks (including those built in the McIntosh period) before carrying on to the side tanks. This, we believe, will give a better and more chronologically-based review.

Lanarkshire Colliery 0-4-0ST, LMS No. 16000
This engine, built in 1896 by Andrew Barclay, was purchased by the Caledonian Railway, in lieu of a debt, and was withdrawn in 1924 without being renumbered. We show it as CR No. 781 in *Plate 121*.

Plate 121 Andrew Barclay 0-4-0ST (allocated LMS No. 16000) in Caledonian colours as it was when CR No. 781. This artist's impression does not indicate any running number.
Authors' Collection

Pre-Drummond 0-4-0ST (LMS Nos. 16001-7; unclassified, later 0F)
The forerunners of the Drummond 0-4-0 saddle tanks dated from 1873 and 1876 and were built by Dübs and Neilson respectively. The details are as follows:

LMS Nos. 16001-2	446 class, built by Dübs 1873 as 0-4-0T, rebuilt to 0-4-0ST in 1890; 3ft. 6in. wheels
LMS Nos. 16003-7	502 class, built by Neilson 1876-81 as 0-4-0ST from the start. Fourteen were built of which five reached the LMS. These engines had 3ft. 8in. wheels and were all reboilered by Drummond, and in this form were almost indistinguishable from the later series *(below)*.

The first pair are represented in *Plate 122*, by which time they differed but little from the later series. No. 16002 was not renumbered.

Plate 122 Dübs-built No. 16001 (ex-CR No. 1162) was rebuilt to this saddle tank form in 1890. The livery Code is C4. Note the shunter's truck with extra coal supply.
Photomatic

The 502 class engines were the true ancestors of the 'standard' Caledonian 0-4-0ST *(below)*. We show in *Plates 123 & 124* two examples in typical LMS condition. The forward projection of the cabside panel and the shorter saddle are the most obvious differences from the Dübs series. We believe that all five were renumbered, and additional to those illustrated we can confirm No. 16003 repainted as in *Plate 124* (Code C14) and No. 16004 with 14in. figures (Code C15). No. 16005 was the last survivor, being scrapped in 1940. It had been the St. Rollox Works shunter since 1925.

Plate 123 No. 16003 (ex-CR No. 1503) — LMS livery Code C4 — was a Drummond rebuild of the original Neilson type. In this form, it was all but identical to the standard 264 class engines. Note the 'above the footplate' springs.
Authors' Collection

Plate 124 This view shows rebuilt 502 class 0-4-0ST No. 16005 (ex-CR no. 1529) when in service as St. Rollox Works shunter — LMS livery Code C14. The somewhat archaic, indeed almost oriental-looking, cab roof and other cab details reflect its Neilson origin.
Authors' Collection

Drummond 264 class 0-4-0ST (LMS Nos. 16008-39; unclassified, later 0F)

These were the standard 'Caley Pugs' and dated from 1885-90. They were basically the Neilson 1876 design but with Drummond details from the outset. The original order included the two 0-4-2STs for the Killin branch, already considered in the previous chapter.

The 0-4-0STs themselves were built both by Drummond and by McIntosh, the latter examples emerging in small quantities between 1895 and 1908. Ultimately, the class totalled 34 (20 Drummond and 14 McIntosh) and the LMS received all but two of them (both from the Drummond series). Several were not renumbered viz. Nos. 16008/12/4/7-8/ 21/3-4/36. There was no significant difference between the two types except that some of the Drummond series (Nos. 16008-25) still had their original 8-spoke 'webbed' driving wheels *(Plate 125)* while the McIntosh series (Nos. 16026-39) had more modern wheels *(Plate 126)*. However, to add confusion, it is clear that a degree of wheel changing between batches took place, accompanied also, we guess, by some wheel replacement. Thus, the whole situation could go into reverse — *Plates 127 & 128*. On some of them *(see pictures)* the wheel springs were positioned above the footplate, a reversion by McIntosh to the original Neilson arrangement.

In use, even though the extended cab sheets of the Neilson and Drummond engines allowed a little more fuel, it was normal for these 0-4-0STs to trail an extra coal supply in a somewhat archaic four wheel truck *(Plates 122 & 129)*.

Plate 125 Drummond-built 264 class 0-4-0ST No. 16010 (ex-CR No. 268), livery Code C14, had retained its original wheels and springs when this picture was taken.
A. G. Ellis

Plate 126 McIntosh-built No. 16031 (ex-CR No. 622), livery Code C21, displays the later driving wheels and 'above the footplate' springs.
A. G. Ellis

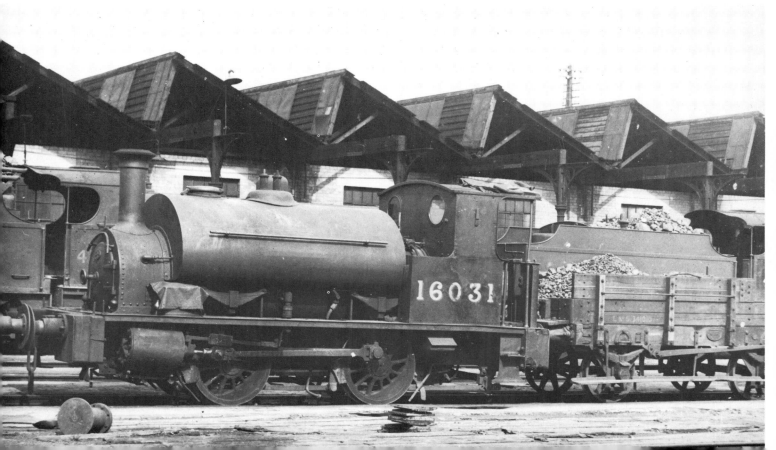

Plates 127 & 128 In these two views, Drummond No. 16025 (ex-CR No. 1515), livery Code C15, has McIntosh wheels with original springs, whereas McIntosh No. 16030 (ex-CR No. 621), livery Code C5, has 'above footplate' springs with Drummond wheels!

A. G. Ellis

Plate 129 (Below) No. 16029 (ex-CR No. 614) seems to carry a version of livery Code C13 or possibly C24, but we can see no signs of the 'LMS'. Note particularly the archaic 4-wheel shunter's truck carrying some pretty gruesome looking coal.

A. G. Ellis

Plate 130 This view shows the St. Rollox livery on No. 16031 (ex-CR No. 622) at Dalry Road in 1932. Note the spark arrestor and the 18in. figures behind the handrail.

L. Hanson

The LMS painted them black and gave them standard liveries. However, at all times there was a degree of indecision as to whether the tank side insignia should go above or below the handrail. The pictures in this section reveal this problem and our summary indicates which option was adopted. There was at least one known example of the St. Rollox variation with 18in. figures *(Plate 130)* but we cannot say if there were any others so finished. Twelve survived into BR days, all but three being of the McIntosh series.

During their lifetime, a few engines, when withdrawn from capital stock, were not immediately scrapped. These were as follows:

LMS No. 16019	Sold to White Moss Colliery in 1935
LMS No. 16025	To service stock at St. Rollox in 1939, replacing No. 16005 *(above)*
BR No. 56027	Withdrawn as Crewe Works pilot — date of transfer to Crewe not known
LMS No. 16037	Sold to Bent Colliery in 1935, later to Stewarts and Lloyd's, in 1945

LMS No. 16025, as BR No. 56025, was given the full BR lining treatment when at St. Rollox Works — and very smart it looked too *(Plate 131)*.

Livery samples are overleaf.

Plate 131 BR No. 56025 still retained its original springs — *see Plate 127* — when in service at St. Rollox in 1953. The engine looks in very splendid condition and also seems to have acquired some 'local' cab sheet modifications.

Authors' Collection

Livery Samples

Code C4 (figures below handrail)	16028/34/7	
Code C5 (figures below handrail)	16030	
Code C5 (figures above handrail)	16019	
Code C14 (LMS above handrail)	16009-10/3/33×/5	
Code C14 (LMS below handrail)	16022	
Code C15 (LMS above handrail)	16020/5/7/34/8	LMS spaced at about 30in. centres in all cases
Code C16 (LMS below handrail)	16011	
Code C21 (LMS above handrail)	16028/31-2	
Code C22 (LMS below handrail)	16020	

× *Base of 'LMS' slightly behind handrail*

0-6-0ST — all types (LMS Nos. 16100-2/16150/16200-29; Power Classes various)

The various residual 0-6-0STs of Caledonian origin fell into several sub-groups, most of them fairly venerable by the start of the LMS period. It seems to us simpler to deal with them together in this survey since, with one exception, they were related types from the Drummond period.

LMS Nos. 16100-2 — These were three survivors of six Drummond 272 class 0-6-0STs built during 1888 as a six-coupled version of the 264 class 0-4-0T *(above)*, and embodying many similar features to the 262 class 'Killin Pug' 0-4-2STs *(see Chapter 2)*. The LMS renumbered Nos. 16100-1 but we can only find clear views of No. 16100 — *Plates 132 & 133*. It was the last to go, in 1929, and the series was unclassified for power.

Plates 132 & 133 Opposite side views of Drummond 272 class 0-6-0ST No. 16100 (ex-CR No. 1273) bearing an almost invisible LMS livery, Code C4.

A. G. Ellis

LMS No. 16150   This was the sole survivor, heavily rebuilt, of the Brittain 486 class 0-6-0ST, dating back to 1881. It was put in LMS Power Class 1. We show it in Caledonian livery in *Plate 134* during 1927, and it was scrapped a year later without renumbering.

Plate 134 The one surviving, much rebuilt, Brittain 486 class 0-6-0ST ran as CR No. 1489 until scrapped, rather than as LMS No. 16150.
A. G. Ellis

LMS Nos. 16200-24   These were 25 survivors of the original thirty members of the Drummond 323 class 0-6-0ST dating from 1887 — the so-called 'Jubilee Pugs'. Most were scrapped before renumbering but seven of them received some form of LMS livery (Nos. 16202/4/11/12/4/20-1). We illustrated No. 16211 in *Plate 189 of Volume One* and gave further examples here in *Plates 135 to 138* with such livery details as we can give on the captions themselves. The class was extinct in 1930 and was put in LMS Power Class 3.

Plate 135 323 class 0-6-0ST, No. 16220 (ex-CR No. 399) seen in LMS livery Code C4. Note the original wheels. Of the other renumbered members of the series, two were painted like this and were structurally similar, viz: 16202 and 16204. The former had new McIntosh wheels.
Authors' Collection

Plate 136 No. 16212 (ex-CR No. 1232) shared its basic anatomy, more or less, with No. 16220 *(Plate 135)* but had 18in. figures, Code C1.

W. Stubbs Collection

Plate 137 A few of the 323 class had extended cab roofs. No. 16214 (ex-CR No. 1235) was one such, and displayed a highly degraded Code C4 livery with no cab panel visible.

Authors' Collection

Plate 138 No. 16221 (ex-CR No. 400) shared its cab roof style with No. 16214 but also had the enclosed back panel. The livery is (just) C4.

A. G. Ellis

LMS Nos. 16225-9   This group of five engines made up the Lambie 211 class, a development of the Drummond 323 class but with Lambie boiler, cab and bunker. The safety-valves over the firebox necessitated a shortening of the saddle tank, and these changes gave quite a different 'look' to the engines. Three were renumbered by the LMS (Nos. 16225/8-9) but we have only confirmed the livery of one of them *(Plate 139)*. We also give a clearer view of one of them in Caledonian colours in *Plate 140*.

Plate 139 Lambie 211 class 0-6-0ST No. 16228 (ex-CR No. 214) in a somewhat scruffy Code C5 livery.
Authors' Collection

Plate 140 Caledonian No. 213 should have become LMS No. 16227 but never did. This picture was taken in LMS days, circa 1926 (note the wagons) and traces of the lined-out CR goods engine livery are clearly visible. Note the spark arrestor on the chimney.

NRM Collection

McIntosh 498 class 0-6-0T (LMS Nos. 16151-73; Power Class 2, later 2F)

Reverting now to the LMS numbering order, the first of the side tank types to be considered are the distinctive short-wheelbase outside cylinder tank engines, introduced in 1912 for working lines of sharp curvature where an 0-4-0ST would be insufficiently powerful. Two only (LMS Nos. 16151-2 — *Plate 141*) were built during the McIntosh period, the rest being built, with little modification to the design, save for larger bunkers and slightly greater rear overhang, during the Pickersgill regime between 1916 and 1922 *(Plate 142)*.

Plate 141 498 class 0-6-0T. No. 16152 (ex-CR No. 1499) was one of only two built in the McIntosh period. The livery is Code C1 but the bunker panel is not visible. We doubt if it was present.
A. G. Ellis

Plate 142 No. 16173 (ex-CR No. 515) was the last of the Pickersgill series of the 498 class and looks very smart in Code C2 livery. The rivetted smokebox and lack of front numberplate make us believe that this was one of the several examples which ran like this well into the 1930s — *see text*.
Authors' Collection

These locomotives were hardly changed during their long lifetime and all passed to BR after the LMS period. In consequence, we can offer a good cross-section of decorative variations for all periods of LMS ownership. The only detail change was the usual change-over to 'pop' safety-valves and the gradual appearance of visibly-rivetted smokeboxes. There was at least one example in late LMS/early BR days fitted with a stovepipe chimney, and one or two more were thus modified in later BR days. *Plates 141 to 145* cover the main LMS liveries and *Plates 146 & 147* the final years. It is, additionally, worth recording that these engines, along with the 29 and 782 class 0-6-0STs *(below)* were quite celebrated for retaining the pre-1928 livery well into the 1930s. It could well have been another example of St. Rollox 'using up' old transfer stocks — somewhat akin to the policy for lined black tank engines *(see Chapter 2)*. The difference is, of course, that in the case of freight tanks, the style was visually indistinguishable from the genuine pre-1928 version, if the red cab panel was employed. We believe that Nos. 16165/70 were probably the last, in 1937 and 1938 respectively. The class was extinct in 1961.

Plate 143 498 class No. 16154 (ex-CR No. 528) displays the somewhat rare Code C4 livery — the only example we know of 14in. figures with this class.

A. G. Ellis

Plate 144 No. 16153 (ex-CR No. 527) represents the common livery for this type during the 1930s — Code C13.

Authors' Collection

Plate 145 (Above) This view of No. 16168 (ex-CR No. 510) shows a common 'late LMS' finish — Code C21 — for this and many other ex-CR classes.

BR (LMR)

Plate 146 (Left) By the end of the LMS period, No. 16154 *(see also Plate 143)* had acquired rivetted smokebox, stovepipe chimney and, we believe, unshaded wartime insignia, Code C24. This picture was taken at Polmadie in 1949.

V. Forster Collection

Plate 147 (Left) Apart from the somewhat uncommon stovepipe chimney, No. 56172 (ex-CR No. 514) was much as it had always been when it went into BR service.

Photomatic

Livery Samples

Code C1	16151/2/6/63/71
Code C2×	16164-7/70/3
Code C4	16154
Code C13	16153/5/8-60/2/9/71-2
Code C21	16163/8-9/72
Code C24	16151-2/4§/6

Notes: × This livery with later cab panel was widespread in the 1930s
§ Stovepipe chimney

McIntosh 29 & 782 class 0-6-0T (LMS Nos. 16230-376; Power Class 3, later 3F)
These were the archetypal Caledonian goods tanks, comprising well over half the total strength of Caledonian freight tanks at the Grouping, and every single one passed to BR.

To be strictly accurate, the class owes its origin to Lambie, who schemed it out before he died. It was left to McIntosh to make the final modifications and supervise their construction in 1895. They were similar in overall dimensions to the Lambie 211 class 0-6-0STs *(above)*, but the large side tanks make them visually very different. The first nine examples (sometimes referred to as the 29 class) were built with condensing apparatus and Westinghouse brakes. The condensing gear was removed during 1920-22 but the Westinghouse pumps remained *(Plate 148)*. This apart, there was no difference between them and the main series — the 782 class — which began construction in 1898 and continued until 1913 *(Plate 149)*. Pickersgill added two final batches in 1916 and 1922 *(Plate 150)*.

Plate 148 29 class 0-6-0T, No. 16232 (ex-CR No. 203) — livery Code C13 — was one of the original condenser-fitted engines, all of which, when reaching the LMS, retained Westinghouse pumps.
A. G. Ellis

Plate 149 The first series of McIntosh steam-braked 782 class 0-6-0Ts had double front footsteps. This is No. 16261 (ex-CR No. 789) of the 1898 batch, livery Code C2.
Authors' Collection

Plate 150 782 class 0-6-0T No. 16360 (ex-CR No. 233) was built during the Pickersgill period in 1916. It carried a somewhat rare livery for the type — pre-1928, Code C4, when at Carlisle in 1924. Note the single front footstep.
Authors' Collection

At the LMS renumbering, the original nine engines were numbered one out of sequence, as Nos. 16231-9, No. 16230 being a 782 class of much later date! However, this was not unusual — *see summary below*. Additionally, although most of the 782 class were steam-braked, a few batches were given Westinghouse equipment. With these engines *(Plate 151)* the reservoir was below the bunker rather than behind the front buffer plank on the 29 class series — *compare Plate 148*. Finally, a few small visible changes, the most obvious being from double to single front footsteps, took place on the batches from late 1898 onwards (LMS Nos. 16230, 16240-53 and 16281 upwards). All these points are summarised below:

LMS No. 16230	782 class, steam-braked (originally Westinghouse) single front footstep, built 1905
LMS Nos. 16231-9	Original 29 class — Westinghouse-fitted, double footstep, originally condensing
LMS Nos. 16240-53	782 class, steam-braked, single footstep, built late 1898/early 1899
LMS No. 16254-80	782 class, steam-braked, double footstep, built early/mid-1898
LMS Nos. 16281-9	782 class, steam-braked, double footstep, built 1904 (No. 16284 in 1910)
LMS Nos. 16290-303	782 class, Westinghouse-fitted, otherwise as Nos. 16281-9
LMS Nos. 16304-46	782 class, as Nos. 16281-9, built 1907-12
LMS No. 16347-51	782 class, Westinghouse-fitted, as Nos. 16290-303, built 1912
LMS Nos. 16352-8	782 class, as Nos. 16281-9, built 1912-13
LMS Nos. 16359-76	782 class, as Nos. 16281-9, built by Pickersgill 1916-22

As stated, the LMS got itself in a bit of a numbering muddle with the 1898-1905 period engines, hence the alternation of details. The Westinghouse-braked examples of the main 782 class series were intended for use in carriage sidings, but as the LMS/BR moved to vacuum braking, some of this series (including No. 16230) lost the brake equipment. We are unable to give a full list of these changes, but during the LMS period, most of the 'fitted' engines remained so-equipped.

Plate 151 No. 16301 (ex-CR No. 647) was a Westinghouse-fitted 782 class locomotive. Note the reservoir below the bunker. It also carries a spark arrestor and, for its type, the somewhat unusual 12in. numerals, Code C14.
A. G. Ellis

Notwithstanding these rather subtle differences, the class in general displayed considerable visual consistency throughout the LMS period. There was the usual change to 'pop' safety valves and visibly-rivetted smokeboxes — fairly consistently applied down the years, but there were relatively few changes in other details. Amongst those which can readily be recognised — and to which we draw attention in the livery samples — are the temporary fitting of spark arrestors to some engines, *e.g. Plates 151 & 152* and the occasional example of a flatter dome cover *Plate 153*. The inevitable stovepipe chimney also made its debut on this class late in LMS days *(Plate 154)* and this became more common in BR days, but was never widespread during the LMS period.

Plate 152 This second view of a spark arrestor, Westinghouse-fitted 0-6-0T No. 16291 (ex-CR No. 637) shows the very common Code C2 livery. Note the brake reservoir just behind the rear footstep, always a visual clue, even if the Westinghouse pump was not visible.

Authors' Collection

Plate 153 No. 16272 (ex-CR No. 800) of the early McIntosh steam-braked series (note the double front footstep) was unusual in being an early example of dome cover replacement, although the steam railmotor in the background indicates that this Code C2 livery was one of the many 1930s' survivals, and that the view (undated) is not as early as it might seem.

Authors' Collection

Plate 154 LMS No. 16266 (ex-CR No. 794) was a distinct oddity, in visual terms. It had a stovepipe chimney and wartime livery (Code C21) much as might be expected, but the retention of the numberplate was extremely unusual at this time.

A. G. Ellis

LMS liveries were conventional throughout. The pre-1928 style was widely adopted and in the round-cornered panel variant (Code C2) remained common through the 1930s for the same reasons given above in respect of the 498 class. We show in *Plate 155*, a picture of what we believe to be the last example to run in this style. The alternative LMS post-1927 styles generally used 10in. figures either plain gold *(Plates 148 & 156)* or yellow with red shading *(Plate 157)*. There were odd exceptions *(e.g. Plates 150 & 151)* but nothing particularly remarkable. The class became extinct in the early 1960s, by which time all were BR black — *Plate 158*.

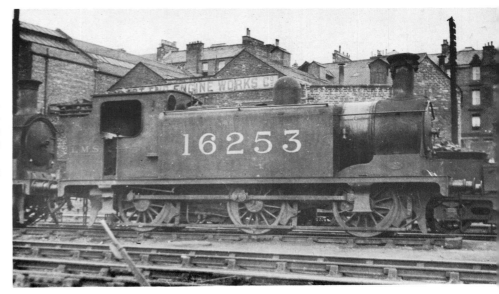

Plate 155 No. 16253 (ex-CR No. 516) was photographed in 1944 still carrying livery Code C2 — the latest example we can find of this livery.

Authors' Collection

Plate 156 This view of No. 16369 (ex-CR No. 251) gives a good impression of a commonly applied livery in the 1930s — Code C13. Note also the supplementary step (and handrail) below the bunker — an unusual but not unique addition to this class, and designed for the human shunter to ride upon.

Authors' Collection

Plate 157 This familiar official view is published again because it gives one of the clearest impressions of the late LMS livery of the 782 class, Code C21. The engine is No. 16342 (ex-CR No. 471).

BR (LMR)

Plate 158 The stovepipe chimney and BR standard livery on No. 56374 (ex-CR No. 394) conclude this pictorial survey of the familiar Caledonian goods and mineral tanks.

A. G. Ellis

Livery Samples

'LMS' spacing post-1927, generally at about 40in. centres

Code C1	16274/347/59
Code C2+	16234-5/37/41/9/53/8/60-1/7/9-70/1*/2×/80*/7/9/91*/2-3/6/300*/2-3/5-6/12-3/6/29/33/41-3/6/52/68/73/5*
Code C4	16251/360
Code C13	16231-3/6/9/42/3/7-8/52/5/7/60/3/6/8/71*/4/83/6/90*/1/9*/308/10/20/2-3/7/32-5/8/45/7/9/56-7/9/69-70/5-6
Code C14	16301*/51
Code C21	16230/3-4/9/47§/57/62-3/6§/75/9§/90/2/300/4/24×/7§/30/5×/42/63-4/6

Notes: + Most of these date from the 1930s — see text
* Fitted spark arrestor
× 'Flat' topped dome
§ Stovepipe chimney

Plate 159 492 class 0-8-0T No. 16501 (ex-CR No. 493, later LMS No. 16951) in pre-1928 livery Code C1 at Motherwell, circa 1925.

A. G. Ellis

McIntosh 492 class 0-8-0T (LMS Nos. 16500-5, later 16950-5; Power Class 4, later 4F)
These heavy duty engines, built in 1903 and 1904, were intended mainly for heavy shunting and banking — and for short distance trains. They were given Westinghouse brakes for operating with the big 30 ton air-braked Caledonian bogie mineral wagons, rather like the similar 0-8-0 tender engines built a year or two earlier — *see Chapter 4*. They were never rebuilt or modified, except for the fitting of 'pop' safety-valves and visibly-rivetted smokeboxes. In 1926 they were renumbered to make room for a large series of new LMS Class 3 standard 0-6-0Ts, and it seems unlikely that all received their original 165XX numbers *(Plate 159)*. We have only traced two, Nos. 16501-2, both livery Code C1.

After the renumbering, three liveries were adopted. First was the pre-1928 style with round panel *(Plate 160)* and thereafter the St. Rollox variation and the conventional post-1927 styles seem to have co-existed *(Plates 161 to 163)*. We have only traced one example which received both forms — *see summary*.

Plate 160 No. 16955 (ex-CR No. 497), after renumbering in 1926. It probably never received its first LMS number (16505), going straight to this form, Code C2.

Authors' Collection

Plates 161 & 162 Opposite side views of No. 16950 (ex-CR No. 492) and 16952 (ex-CR No. 494) show the St. Rollox livery variation on these imposing 0-8-0Ts.
Authors' Collection

Plate 163 (Below) Another view of No. 16955 *(see Plate 160)*. This was the last survivor of the type, and is one of at least three examples known to have received the proper 1928 livery style, Code C15. It was photographed, in August 1937, at Polmadie.
L. Hanson

The LMS kept them going for quite a long time, given their small numbers and somewhat non-standard nature. They had a boiler (and other fittings) which was based on that of the Drummond 0-6-0s, so this may have helped a little. Withdrawal began in 1932 and the last, No. 16955, was scrapped early in 1939.

Livery Samples

Code C1   16501-2
Code C2   16953-5
Code C3   16950-2 (in the St. Rollox style — i.e. 14in. LMS on bunker)
Code C15 16952/4-5 (letter spacing about 40in.)

Tailpiece This sort of view captures, in a nutshell, many typical aspects of the LMS scene north of the border. The engine is Drummond 264 class 0-4-0ST No. 16011 (ex-CR No. 270), but has post-grouping safety-valves; the Caledonian shunters' truck has LMS livery; there is a fairly modern tank wagon and the picture was taken at Inverness, of all places, in August 1939.

L. Hanson

Chapter 4
Caledonian Railway — Freight Tender Classes

This group of engines represented some 42 per cent, approximately, of the total Caledonian fleet at the Grouping and most of them were 0-6-0s. Of the 0-6-0s, all but two derived from Drummond's pioneer design so, in the freight tender group, the Drummond influence was even stronger than in the categories considered so far.

Essentially, the collection in 1923 amounted to a handful of pre-Drummond engines, mostly 0-4-2s, two very big series of 0-6-0s (very much related to each other) and another group of somewhat specialised engines in the 2-6-0, 4-6-0 and 0-8-0 wheel arrangements. As usual, we deal with them in ascending LMS number order.

Brittain 670 class 0-4-2 (LMS Nos. 17000-20; Power Class 1)
These quaint-looking engines were the survivors of a series of 30 such machines dating from 1878, and they had a visual character all their own. By the time the LMS received them, scrapping had commenced, and not much more than half of the 21 residual members received their LMS numbers, viz: Nos. 17002-5/7-8/11/13-4/17-20. From 1904 onwards, all had received new boilers and standard 'Caley' chimneys. They had also been given Westinghouse brake equipment for passenger working. Tender styles were somewhat variable and we give some details in the picture captions.

We presume that all those renumbered were given the pre-1928 livery as in *Plates 164 to 167*, but we cannot confirm all of them. However, Nos. 17003-5/7/14 were similar to those illustrated. No. 17018 had the round-cornered cab panel with 14in. figures at first, and then received 18in. figures — *Plate 168*. We cannot say if any others were similar in either respect. However, for a while, No. 17000 ran as CR No. 1672, with a repainted tender bearing the number 17005!

Although the last survivor lasted until 1932 (No. 17003) we have no trace of any of them bearing a post-1927 style livery.

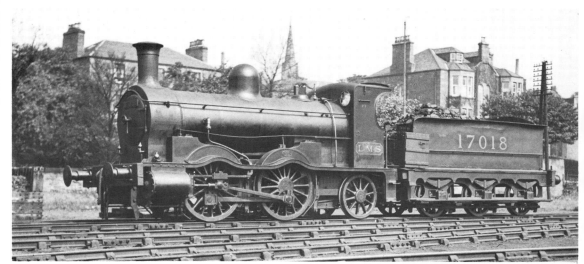

Plates 164 & 165 Opposite side views of Brittain 0-4-2 locomotives Nos. 17002 (ex-CR No. 1676) and 17018 (ex-CR No. 1275), both in pre-1928 LMS livery, Code C4. Note the much older tender attached to No. 17002, believed to be from an earlier 2-4-0. No. 17018 is attached to the most common variety of tender associated with the class in LMS days.
BR (LMR) and H. C. Casserley

Plates 166 & 167 Two more tender varieties are shown in these less than perfect views of Nos. 17011 (ex-CR No. 165) and 17020 (ex-CR No. 1717), both in black livery, Code C4. At least two other examples (Nos. 17005/8) had four wheel tenders like that of No. 17020 in LMS days, but that of No. 17011 is thought to have been unique, although somewhat similar to the version in *Plate 168*.
W. T. Stubbs and Authors' Collection

Plate 168 (Below) No. 17013 (ex-CR No. 279) was fitted to what we believe to have been the only tender of this precise type associated with the 0-4-2s and which also carried 18in. figures — Code C2. At an earlier date, its tender was like that of No. 17018 *(Plate 165)* as were the tender-side numerals.

Authors' Collection

Solway Junction 0-6-0 (LMS Nos. 17101-2)
These two engines came from the old Solway Junction Railway and dated from 1868. They were conventional 0-6-0s of broadly identical nature, except for their tenders, and both were renumbered. They survived only until 1927 and we illustrate them in *Plates 169 & 170*.

Plates 169 & 170 The two ex-Solway Junction 0-6-0s, Nos. 17101/2 (ex-CR Nos. 381/2), both seen bearing livery Code C4, were, in theory, identical. By LMS times they had acquired different style tenders to each other and No. 17102 had two pairs of coupled wheels which looked a little as though they had escaped from a contemporary Bassett-Lowke model locomotive catalogue!

*A. G. Ellis
and Authors' Collection*

Drummond 294 class 0-6-0 (LMS Nos. 17230-473; Power Class 2, later 2F)
These engines, the famous 'Caley Jumbos', were designed by Drummond in 1883 and built continuously thereafter into the McIntosh period until almost the end of 1897, by which time they had become by far the most numerous single Caledonian type. Even by LMS days they still represented almost 25 per cent of the capital stock, considerably outnumbering their larger successors *(below)*.

Although, essentially, the whole series was basically an 0-6-0 version of the 66 class 4-4-0s, they can, conveniently, be divided into three sub-groups for analysis, essentially reflecting the periods of office of the three engineers who built them.

a) Drummond series — Generally all the same but given a variety of different tenders depending on the time when built. These became LMS Nos. 17230-351 plus No. 17357. A small number were Westinghouse-fitted (LMS Nos. 17319-24 and 17342-7).

b) Lambie series — A slight modification with Lambie boilers having safety-valves over the firebox and the dome set further forward. These became LMS Nos. 17352-6/8-86/9-92. No. 17357 got itself mixed into this series in 1923 (it was a Drummond engine) and Nos. 17387-8, although part of a Lambie order, were delivered to the McIntosh specification *(overleaf)*. A few of these, like the Drummonds, were Westinghouse-fitted (LMS Nos. 17367-76).

c) McIntosh series — A further build from 1895 onwards, sometimes referred to as the 709 or 711 class, depending on whether the two engines ordered in the Lambie period (CR Nos. 709-10) are counted in. The LMS numbers for this series were 17387-8/93-473. All were Westinghouse-fitted and classified 'mixed traffic', and this was the main change from most of the earlier ones.

In point of fact, there was very little visual difference between any of them in the early days, except for the safety-valves on the domes of the Drummond series. Very few came to the LMS in the latter form *(Plate 171)* and by the post-grouping time, the engines themselves were a fairly homogenous group, quite well typified by *Plate 172*. Most of the Drummond series had, by then, received replacement Lambie pattern boilers. Apart from safety-valves, the original boilers could also be identified by the two wide boiler cladding plates between firebox and smokebox rather than the three plates of the Lambie-McIntosh boiler. Subsequent boiler changing sometimes put earlier pattern boiler clothing on post-Drummond engines — *e.g. Plate 173* — it being found that by simply reversing the second section of clothing, the opening for the dome was in the appropriate position for the replacement boiler. The situation, therefore, could be quite confusing.

Plate 171 Drummond 0-6-0 No. 17269 (ex-CR No. 1517) was one of only a few 'Jumbos' which retained the original styling of both engine and tender into LMS days. Note the dome-mounted safety-valves, the dome position central above the driving axle, and the original Drummond tender with underhung springs — type D1. The livery code is C14.
Authors' Collection

Plate 172 No. 17357 (ex-CR No. 550) has a Lambie style '3-plate' boiler with conventional dome set somewhat forward of the driving wheel centre line — probably fitted after its accident at Pollockshaws in 1923. The tender is type D3 from a withdrawn 66 class 4-4-0 but, this apart, the locomotive configuration is probably the most common form in the LMS period. The livery code is C14.
Authors' Collection

Plate 173 No. 17366 (ex-CR No. 563), livery Code C14, was a post-Drummond example but carrying '2-plate' boiler cladding when photographed at Dawsholm in 1935, coupled to a type M4 tender from a withdrawn 55 class 4-6-0.
A. G. Ellis

Subsequently, 'pop' safety-valves replaced the old type and some engines received flatter-topped dome covers; which revealed LMS built boilers with changed dome construction. Additionally, one or two sported tender cabs. *Plate 174* displays a particularly interesting example embodying most of the features mentioned so far. Finally, in late LMS/early BR days, some engines began to sport stovepipe chimneys of not particularly pleasing appearance, *(Plate 175)* frequently with some of the other changes too. Thus, as the years went by, individual differences began to mount up between different members of the class.

Plate 174 'Jumbo' No. 17322 (ex-CR No. 413), livery Code C21, sported a '2-plate', LMS built boiler (with flat-topped dome) and Lambie type L2 tender fitted with a tender cab when this picture was taken, probably during the war.
A. G. Ellis

Plate 175 This particular variation shows No. 17292 (ex-CR No. 1318), livery Code C21, towards the end of LMS ownership with '3-plate' boiler, flat-topped dome, stovepipe chimney and replacement type M2 tender.
Authors' Collection

This variation, however, was nothing compared with the tender changing which took place over the years. We have already explained *(in our introduction to Chapter 1)* how the Caledonian built a bewilderingly large number of differently-styled tenders, and the 'Jumbos' undoubtedly held the prize for the number of different types with which they ran. We have identified at least nine different varieties so this seems to be a good point at which to recapitulate the story as far as the six wheel tenders on the 'Jumbos' are concerned. Readers should consult *Table 1, page 6,* for more specific details.

The original Drummond tender (type D1) was based on the Stroudley LB&SCR pattern, and was fairly consistently used from 1883-9. It gained a reputation for poor riding with its underhung springs, but many still survived into the LMS period *(Plate 171)*. The next two Drummond tenders (types D2 and D3) were built for passenger engines but, when these were scrapped, the LMS put many of them behind 'Jumbos' to replace D1s. Examples of each are in *Plate 176* (D2) and in *Plate 172* (D3).

Plates 176 (Bottom Left) & 177 to 179 These four pictures although showing further locomotive detail and livery variations within the 294 class, have been chosen particularly to illustrate four more tender varieties. The details are as follows: No. 17230 (ex-CR No. 259), livery Code C1, tender type D2; No. 57459 (ex-CR No. 595, LMS No. 17459), early BR livery, tender type D4; No. 17320 (ex-CR No. 411), livery Code C2, tender type L1; No. 17330 (ex-CR No. 378), livery Code C14, tender type P2.

A. G. Ellis and Authors' Collection

From 1889, the second 'standard' tender for the new 'Jumbos' was the fourth Drummond variant (type D4), and a considerable number of these were to be seen with the 0-6-0s — e.g. *Plate 177*.

Withdrawal of further 4-4-0s caused some of the not very numerous Lambie 3,570 gallon tenders (type L1) to become spare, and these, in turn, replaced more of the D1 type on older engines *(Plate 178)*. However, long before this took place, the final series of 'Jumbos' had been built with the second variety of Lambie tender (type L2), designed specifically for this class of engine. All the McIntosh-built engines came into service with this type, and most retained them into LMS days — *Plate 174*.

Further withdrawals of other ex-CR engines during the LMS period caused examples of three more varieties of tender to be available to replace some of the ageing D1 types on the 'Jumbos'. These were the standard McIntosh 3,000 gallon tenders (type M2 — *Plate 175*) the short wheelbase McIntosh 3,000 gallon tenders (type M4 — *Plate 173*) and the standard Pickersgill 3,000 gallon tenders (type P2 — *Plate 179*). *Table 1* gives the classes from which these tenders came.

Given all the possible permutations of locomotive variations and tender styles within this class during LMS days, we feel safe in asserting that, along with the Midland 0-6-0s *(Volume Four)*, the Caledonian 'Jumbos' are likely to cause more problems of detail verification — especially to modellers — than any other groups we shall cover. For this reason, we have given this extended analysis here, and our summaries *(below)* try to cope with most of the variations we have identified.

Fortunately, when it comes to LMS liveries, these engines were very consistently painted, whatever their external detail differences might have been. Prior to 1928, the proper freight livery was universally adopted with few known 'jokers'. Both numeral heights were used, usually depending on the tender type, and both varieties of cab panel were employed, the change in panel style being circa 1926.

After 1927, an astonishing degree of consistency was to be seen with the well-nigh universal use of 12in. figures, 53in. LMS spacing, and plain gold characters. Very few exceptions to this are known to us and *Plate 180* probably suggests why! From the very late 1930s, an equally consistent use was made of the red shaded yellow characters with 10in. numerals. Obviously there were a few exceptions, but not many, and our lists below illustrate this point.

Plate 180 No. 17351 (ex-CR No. 1553) is the only 'Jumbo' we have managed to confirm with 14in. figures, Code C15. Anything other than a '1' as the last digit would have caused a few problems in the paintshop, we reckon! Note the original boiler and safety-valves.

A. G. Ellis

All but a handful of these superb little engines reached BR, and gave many more years of service, before the last few vanished in the early 1960s. It is a thousand pities that one was not preserved — they earned their keep far more than did many more glamorous types which have survived.

We conclude this section with as comprehensive a summary as we can, including a few additional views to show more variations *(Plates 181 to 186)*.

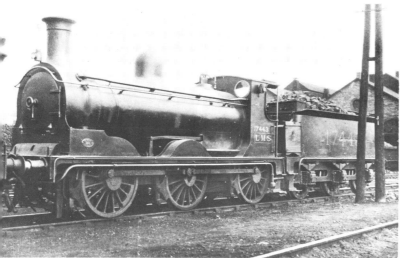

Plate 181 (Above) The McIntosh-built 'Jumbos', sometimes known as the 709 class to differentiate them from the earlier series, were, in LMS days, probably more visually consistent than the older engines of the type, particularly in the matter of tenders — type L2 for the most part. A nice ex-works example is seen here, No. 17468 (ex-CR No. 765) in pre-1928 livery, Code C1.
Authors' Collection

Plate 182 (Left) No. 17443 (ex-CR No. 574), livery Code C2, was given a cabside number as well, probably about the time of the livery change.
Authors' Collection

Plate 183 (Right) A pleasing combination of original Drummond boiler, tender cab and first LMS livery, Code C2, was displayed by No. 17371 (ex-CR No. 701) at Dawsholm in the early 1930s. This engine, when new, had the later pattern boiler, not the version illustrated.
A. G. Ellis

Plate 184 Flat dome and first style Drummond tender distinguished No. 17244 (ex-CR No. 403) when seen at Stirling in 1938, Code C14.
A. G. Ellis

Plate 185 (Below) No. 17366 *(see also Plate 173)* still carried its type M4 tender at Balornock in 1950, and also its post-war LMS livery, Code C21, but it now has a flat dome.
A. G. Ellis

Plate 186 A pleasing view of No. M17240 (ex-CR No. 352) in early BR livery concludes this review of Drummond 'Jumbos', even though the possible permutations are by no means exhausted! Note the flat dome and type D4 tender.
A. G. Ellis

Livery Samples and detail variations

Preliminary Notes

1. Tender type is given in brackets after engine number — using the reference numbers from *Table 1, page 6*
2. Assume '3-plate' boiler lagging, normal dome and Caledonian chimney unless indicated otherwise
3. After the tender reference, the following suffixes are used for detail variation, if applicable:
    a) Drummond boiler with safety-valves on dome
    b) Lambie/McIntosh boiler with two lagging plates and, of course, safety-valves on firebox
    c) 'Flatter'-topped dome — LMS built boiler
    d) Stovepipe chimney
    e) Tender cab
4. Letter spacing 1928-47 was normally about 53in. between centres

Code C1   17230(D2)/17244(D1)/17248(D1)/17250(D4)(b)/17254(D1)/17373(L2)/17414(L2)(b)/17460(L2)(e)/
           17468(L2)/17470(L2)
Code C2   17237(D4)(a)/17285(D1)/17286(D1)(b)/17289(D4)(b)/17296(D2)/17320(L1)/17330(L2)(b)/17334(D1)(a)/
           17335(D4)(b)/17357(D3)/17371(L2)(a)(e)/17385(D4)/17393(L2)/17401(L2)/17430(L2)/17433(L2)
           17439(L2)/17440(L2)/17443(L2)(b)/17455(L2)
Code C4   17251(D1)(b)/17287(D1)/17311(D1)(a)/17349(D4)(a)17356(L2)/17360(D4)/17390(D1)/17416(L2)
Code C13   17300(M4)(c)/17310(D4)/17338(L2)
Code C14   17232(D4)/17244(D1)(c)/17245(D4)/17246(D4)(c)17252(D4)(c)/17257(D4)(b)/17264(D4)17265(L1)(c)/
           17269(D1)(a)/17272(D1)/17273(D4)/17275(D1)(c)/17279(D4)(b)/17282(D1)/17286(D4)(b)/17294(D1)(b)/
           17295(D3)/17296(D2)/17297(D1)/17298(D1)/17299(D4)/17301(D1)/17302(D4)/17303(D1)(c)/17307(P2)
           17311(D1)(b)/17314(D2)/17318(D2)/17324(L2)/17326(D4)(b)/17328(L2)/17330(P2)(b)/17332 (D1)(b)/
           17333(D4)(b)/17337(L2)/17340(D4)(b)/17342(L2)(c)/17343(D4)(c)/17344(L2)(b)/17347(D4)/
           17349(D4)(b)/17351(L2)(b)/17352(D4)(b)/17353(D1)/17357(D3)/17358(D4)(b)/17359(L2)/
           17362(D4)(b)(c)/17363(D4)/17366(M4)(b)/17369(L2)(b)(c)/17377(D2)/17378(D1)(c)/17385(D4)(c)/
           17388(L2)(c)/17389(D1)/17391(D1)/17392(L2)/17393(L2)(c)/17398(L2)/17400(L2)/17402(L2)(e)/
           17403(L2)/17404(D4)(c)/17405(L2)(c)(e)/17406(D4)/17408(L2)(a)/17409(L2)/17410(L2)/17413(D4)/
           17415-9(L2)/17422-3(L2)(c)/17430-1(L2)/17435(L2)(c)/17437(D4)/17438(L2)(c)/17440(L2)(e)/
           17442(L2)(c)/17444(L2)/17446(L2)(c)/17447-8(L2)/17451(L2)(c)/17452(L2)/17456(L2)/17459-60(L2)(c)/
           17461(L2)/17465-7(L2)/17470(L2)/17471(L2)(e)
Code C15   17351(L2)(a)
Code C21   17233(P2)(c)/17239(D1)(d)/17249(D4)(b)/17254(D2)(c)(d)/17258(M2)(c)/17259(D4)(c)/17266(D4)/
           17270(M2)(c)/17292(M2)(c)/17305(D1)(b)/17306(M4)(b)/17316(L2)(c)/17320(L2)(c)/17322(L2)(b)(c)/
           17336(D4)/17339(P2)(b)/17350(D2)/17359(L2)(c)/17366(M4)(b)(c)/17372(D4)(c)(e)/17380(P2)/
           17383(D1)/17396(L2)/17402(L2)(c)/17411(L2)/17415(L2)(d)/17424-5(L2)/17436(L2)(c)/
           17440(L2)(c)(d)(e) /17443(L2)/17449(L2)(c)/17454(L2)(d)/17455(L2)/17457(L2)/17468(L2)

NB Some of the Code C21 examples may have been Code C24 — wartime sources are not always 100 per cent clearly defined

McIntosh 812 class 0-6-0 (LMS Nos. 17550-645; Power Class 3, later 3F)

After the complexities of the 'Jumbos', it is something of a relief to turn to the relative simplicity of their larger-boilered and more powerful successors. It would be possible to deal with them all together but they did form three, readily identifiable, sub-groups, separately identified by class designation in Caledonian days — and this is how we deal with them here.

The McIntosh 812 class was essentially an enlarged 'Jumbo' with larger boiler and cylinders. The boiler was of the 'Dunalastair I' type, the cab style was originally, 'Dunalastair II' inspired, and they first emerged in 1899. The last batch to be built, in 1908-9 (LMS Nos. 17629-45) saw a change in cab shape to the 'Dunalastair III' pattern (with a continuous curve to the cut-away) and deeper main frames behind the splashers. A new standard 3,000 gallon tender was fitted to the engines (type M2 — *Table 1*) and very little tender changing took place. The two basic variants are shown at *Plates 187 & 188*.

The first seventeen (LMS Nos. 17550-66) were Westinghouse-fitted and regarded as 'mixed traffic' *(Plate 189)* but the rest were steam-braked only. In LMS days, a few of these later ones were modified *(e.g. Plate 190)* but we are unable to give a full list or say how many were thus treated — very few we guess. The LMS never really changed these engines, apart from substituting 'pop' safety-valves, and they mostly served until BR before major withdrawal commenced. The exceptions were Nos. 17567/98/610, withdrawn during 1946/7. CR No. 828 (LMS No. 17566) of the Westinghouse series is preserved on the Strathspey Railway.

Plates 187 & 188 These two pictures show the original version of the 812 class, with No. 17571 (ex-CR No. 833) in post-1927 livery, Code C15, and the later version No. 17632 (ex-CR No. 855) in the early livery, Code C1. Note the difference in shape between the cabside 'cut-away', also the retention of Ramsbottom pattern safety-valves on the older engine.
A. G. Ellis

Plate 189 No. 17553 (ex-CR No. 815), livery Code C1, was one of the Westinghouse-fitted examples, and is seen here in passenger service near Ibrox on the Wemyss Bay Service, soon after the Grouping.
Authors' Collection

Plate 190 (Below) No. 17618 (ex-CR No. 283) was one of a few engines vacuum-fitted by the LMS after the Grouping. This picture is also particularly useful in showing clearly the 'highlight' lines on the 14in. black shaded Midland pattern numerals, livery Code C15, used on quite a few of these engines.
A. G. Ellis

Plate 191 This close-up view of No. 17578 (ex-CR No. 840), photographed at Dalry Road in June 1936, clearly reveals the black shaded 12in. numerals on a black engine — Code C14. Paradoxically, this feature showed up rather better on a dirty engine than on a clean one!
Authors' Collection

In terms of livery, there was considerable pre-1928 consistency amongst those repainted and we give a good sample below. After the change of style in 1928, both 14in. and 12in. numerals seem to have been represented, the latter being somewhat more numerous. We think that the 14in. numerals were probably more popular in the early 1930s (probably to use up old pre-1928 stock) and that later on a change to 12in. took place. This seems to have been quite common in Scotland — somewhat analogous to using up the old 18in. figures on tank engines *(Chapter 2)* but beyond that it would be foolish to speculate. During this phase, insignia were gold with black shading (which showed up as plain) and this is clearly seen on some examples — *e.g. Plate 191.*

The usual Scottish area change to 10in. figures took place in the late 1930s and 1940s *(Plate 192)*, and there is the usual problem of deciding whether wartime repaints employed plain yellow or shaded yellow characters. We give our 'best' estimate in the summary below.

Finally, one or two of the 812 class locomotives received replacement tenders of different style *(e.g. Plates 193 & 194)* and a few received a more flat-topped dome cover in later years *(Plate 195)*. Our summaries try to take account of these small variations.

Plate 192 This 812 class locomotive, No. 17645 (ex-CR No. 661) was from the later series (note the cabside shape) and is seen here in the Edinburgh area in 1947 with 10in. cabside numerals — Code C21.
Authors' Collection

Plate 193 (Above) No. 17552 (ex-CR No. 814), livery Code C24, was one of but a few 812 class engines to receive a replacement tender — in this case type M3, probably from a withdrawn 4-6-0 or 0-8-0 *(see Table 1, Chapter 1)*. Most of these tenders went to displace bogie types on the 4-4-0s, but this picture clearly proves that this was not always the case.
A. G. Ellis

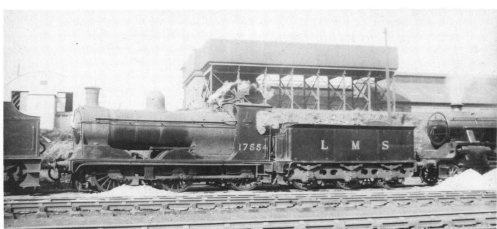

Plate 194 (Above) No. 17554 (ex-CR No. 814), livery Code C24, sported a very unusual tender for this class when photographed, circa 1945. It was a type M1 tender *(see Table 1, Chapter 1)* from a withdrawn 'Dunalastair I', and we know of no other 812 class engine with this pairing.
A. G. Ellis

Plate 195 (Right) This 812 class engine, No. 17611 (ex-CR No. 873), livery Code C21, displayed both a replacement Pickersgill type P2 tender and a flat-topped dome when photographed in rather grubby condition after World War II.
Authors' Collection

Livery Samples

Assume type M2 tender (unless indicated) and 53in. letter centre spacing, post-1927

Code C1	17551/3-4/62/9/71/604/6/23/5/30/2-3/43
Code C2	17572/4/9/81/3/40
Code C14	17557-8/63-5/7/70/2×/4/6/8/80-2/4/90-1/9×/600§/4*/6/16-7/22-3/6-7/35-6/8-9/44
Code C15	17551/71/608/10/11×/3×/5/8/21/31×
Code C21	17553×/8×/66/77/93×*/5-6×/604-5/9×/11×*/20/4×/5/45
Code C24	17552§/4+/61/75§/607×

Notes: × *'Flat' topped dome*
  § *type M3 tender*
  * *type P2 tender*
  + *type M1 tender*

McIntosh 30 class 0-6-0 (LMS Nos. 17646-9; Power Class 3, later 3F)
In 1912, following the success of superheating on the 4-4-0s and 4-6-0s, McIntosh built a superheated version of the 812 class with piston valves above the cylinders — the 30 class. Only four were built, and their much heavier front end and higher-pitched boilers gave them distinctive and, for a McIntosh engine, somewhat ungainly lines *(Plate 196)*. Their heavy front end obviously caused McIntosh to re-think the design because, later in 1912, the next batch of five were turned out as 2-6-0s *(below)*. However, the superheating experiment with this type of engine ended with the 30 class (and, of course, the subsequent 2-6-0s).

Plate 196 This view of 30 class 0-6-0 No. 17648 (ex-CR No. 32), livery Code C14, clearly shows the somewhat altered 'front-end' of the superheated McIntosh 0-6-0s. The engine was photographed at Inverness in 1931, a somewhat unusual location this early in LMS days.

G. Coltas

Plate 197 An opposite side view of superheated No. 17646 (ex-CR No. 30), livery Code C14, further emphasises the less than harmonious looks of the few 30 class 0-6-0s.

Authors' Collection

We believe the LMS painted them all in the pre-1928 livery but can only confirm No. 17646 (Code C2). Subsequently, all four received the 12in. numerals as depicted in *Plates 196 & 197*, Code C14, but we have no other details. None survived the LMS, No. 17646 being scrapped in 1936, the other three during 1945/6. All four ran with type M2 tenders throughout.

Pickersgill 300 class 0-6-0 (LMS Nos. 17650-92; Power Class 3, later 3F)
At the end of World War I, Pickersgill introduced his own composite version of the 812 and 30 class 0-6-0s, reverting to the non-superheated form. Visually and dimensionally they were all but identical to the McIntosh engines. However, the shorter smokebox and Pickersgill chimney gave an altered 'front end' look, and the tenders (type P2) had coal rails to distinguish them from the otherwise very similar type M2 tenders of the 812 and 30 classes. The piston valves and higher-pitched boilers of the 30 class were married to the somewhat smaller cylinders of the 812 class to produce a robust and reliable type. Some few had balanced slide valves but without significant change of external lines.

The LMS superheated most of them, but otherwise made hardly any changes, except for the gradual substitution of 'pop' safety-valves and visibly rivetted smokebox wrappers as years went by. Liveries were orthodox, and we give a fair cross-section in our summary and in *Plates 198 to 201*.

Proportionally, they did not last quite as well as the 812 class, withdrawal commencing in the 1930s, including all the residual saturated examples. By BR times quite a number had gone, viz. Nos. 17656-7/60/2/4/75-8/80/3/5/92, most of them before the war. It is reasonable to assume that the war delayed withdrawal of quite a number of them and, in the event, many served until the early 1960s.

Plates 198 & 199 Opposite side views of Nos. 17681 (ex-CR No. 280) and 17688 (ex-CR No. 675), livery Codes C1 and C14 respectively, reveal the well-proportioned lines of the Pickersgill 300 class 0-6-0s — the last Caledonian engines of this wheel arrangement.

Authors' Collection and A. G. Ellis

Plate 200 (Below) This view of No. 17652 (ex-CR No. 296) gives a clear indication of the piston valve covers and visibly rivetted smokebox of a Pickersgill 300 class 0-6-0 in late LMS days — livery Code C21.

A. G. Ellis

Plate 201 No. 57686 (ex-CR No. 673) reveals its piston valves even more clearly in this view, photographed in May 1949. The tender still retains pre-1948 livery (probably Code C21) and the engine carries an experimental shape of BR numerals, not uncommon in Scotland.

A. G. Ellis

Livery Samples

Assume type P2 tender unless indicated

Code C1	17667/77/81/4/90
Code C2	17657/9/62-3/8/78-9/85
Code C4	17650
Code C14	17650/60/2/6/9/70-6/82-4/8-9/92 ×
Code C15	17651/8/79
Code C21	17652-3/9/91

} letter spacing at about 53in. centres.

× Also ran in this livery with type M2 tender later

McIntosh 34 class 2-6-0 (LMS Nos. 17800-4; Power Class 3, later 3F)

As mentioned above, the McIntosh superheated 0-6-0s of the 30 class turned out to be a little heavy at the front end, and the next series of five engines was given a pair of leading wheels to help spread the load. They were the first Scottish 2-6-0s, but the extra pair of wheels always gave the appearance of having been added as an afterthought — as indeed they had — *Plates 202 to 204*. They were, wheel arrangement excepted, identical to the 30 class, and were paired with the same type tenders (M2).

Plate 202 No. 17804 (ex-CR No. 38) was the last of the 34 class 2-6-0s, and is seen in early LMS livery, Code C2. One has almost to look twice before the leading wheels become 'part' of the ensemble!

A. G. Ellis

Plate 203 A rear view of No. 17801 (ex-CR No. 35), livery Code C1, shows good detail of the cab interior and the type M2 tender, and again shows the 'supplementary' nature of the leading wheels.
A. G. Ellis

Plate 204 2-6-0 No. 17802 (ex-CR No. 36), livery Code C14, gives a very clear impression of the extended 'front end' of these somewhat curious 34 class 2-6-0s.
W. L. Good

Interestingly the G&SWR had a somewhat similar problem with the Peter Drummond 0-6-0s and also went for a 2-6-0 continuation — but with rather better visual results *(Chapter 6)*.

In LMS days, no significant changes took place to them, and all were withdrawn during 1935-7. We believe that all carried both the pre-1928 and post-1927 liveries, but we only give confirmed examples.

Livery Samples

Code C1   17801
Code C2   17803-4
Code C14   17802-3 } letter spacing at about 53in. centres.
Code C15   17800-1

McIntosh 918 class 4-6-0 (LMS Nos. 17900-4; Power Class 3, later 3F)
In *Chapter 1*, we outlined the somewhat tentative flirtation with the 4-6-0 type during the McIntosh and Pickersgill periods, and also described the evolution of the engines which the LMS classified as 'passenger' types. We now turn to the two further examples of this wheel arrangement which, because of their regular duties, the LMS chose to put in the freight lists, starting with the small-wheeled 918 class.

These engines were, in essence, a large-boilered version of the 55 class *(Chapter 1)* embodying identical mechanical components, but a shortened version of the 903 'Cardean' class boiler. They were built in 1906 to work express goods traffic and this may have prompted the LMS to classify them as 'freight' — but the 'Caley' had painted them blue and also used them on occasional passenger trains; so they were in all respects genuinely 'mixed traffic' types. They always ran paired with 3,570 gallon type M3 tenders *(Table 1 — Chapter 1)*.

During LMS days, smokebox wingplates were removed and 'pop' safety-valves substituted, but otherwise little change took place. All were repainted in the standard pre-1928 black livery (Code C1) shown in *Plate 205* and none, we believe, received the later style markings. All were withdrawn during 1929/30 but the boilers from Nos. 17901/4 were used for the rebuilding of the two 55 class engines mentioned on *page 32*, which in effect made the engines 'quasi-918 class' after the real type was extinct!

Plate 205 The first of the 918 class, ex-CR No. 918 itself, is pictured ex-works in the pre-1928 LMS livery — Code C1. This is the only known decorative style given to this series of engines during LMS days, as far as we can establish.
Authors' Collection

McIntosh 179 class 4-6-0 (LMS Nos. 17905-15; Power Class 3, later 3F)

If the LMS decision to put the 918 class into the freight lists was somewhat odd, then the similar exercise with the 179 class was more so — for they were simply a superheated version of the 908 'Sir James King' class, embodying the more modern 'two window' cab of the last of the saturated series — *see page 33*.

It is true that the engines were intended for and spent most of their time on express goods traffic — but then so too, inter alia did the passenger classified 908s! Pre-1923, all were painted passenger blue and, for the record, the 179 class were the last McIntosh engines to appear before he retired. They were built in 1913 and 1914 in quantities of five and six respectively. They were typically well-proportioned McIntosh engines *(Plates 206 to 208)* and always ran with 3,570 gallon type M3 tenders *(Chapter 1 — Table 1)*.

Plates 206 & 207 These two views of 179 class engines Nos. 17906 (ex-CR No. 180) and 17909 (ex-CR No. 183) in LMS livery Code C1, nicely show the well-balanced lines of this superheated 5ft. 9in. 4-6-0 type.
A. G. Ellis and BR (LMR)

Plate 208 This semi-close-up-left-hand side view of No. 17914 (ex-CR No. 188) shows an engine with the later pre-1928 cab panel, Code C2, and replacement 'pop' safety-valves.
Photomatic

In LMS days, the usual removal of smokebox wingplates took place along with replacement of 'pop' safety-valves, but nothing of any real consequence was changed and they lasted longer than any other McIntosh 4-6-0s. Withdrawal did not start until 1935 and two survived the war, No. 17908 *(Plate 210)* being the last survivor (1946). Liveries ran the whole gamut of styles, except that we have no record of 12in. cabside numerals being used after 1927. This may be because the official edict was to use the largest figures possible, and the 179 class had bigger cabside panels than other ex-Caledonian engines so could readily accept the 14in. numerals without undue difficulty *(Plate 209)*.

Livery samples overleaf

Plate 209 No. 17911 (ex-CR No. 185) displays the common post-1927 condition of the 179 class — livery Code C15.
Photomatic

Plate 210 No. 17908 (ex-CR No. 182) was the last McIntosh 4-6-0 to survive — livery Code C21. This picture was taken at Corkerhill in 1945.
Authors' Collection

Livery Samples

Code C1	17905-6/9/12/5
Code C2	17907/8/14
Code C15	17905/9-13
Code C21	17908

McIntosh 600 class 0-8-0 (LMS Nos. 17990-7; Power Class 6)

These eight engines were built, two in 1901 and six in 1903, to pull the heavy coal trains in the Lanarkshire Coalfield area. They were associated with the 30 ton bogie coal wagons introduced at the same time by McIntosh, and both engines and wagons were Westinghouse-braked — a surprisingly modern concept for the early part of this century. They proved a little embarassing in that their haulage capability exceeded the contemporary siding lengths in many areas — shades of the Gresley 2-8-2s of later years on the LNER!

Visually, they were typically McIntosh, but with extremely long boilers, basically a lengthened version of that used on the 'Dunalastair III' series. Above the footplate they were as good looking as any other British 0-8-0, but their unequally divided coupled wheelbase gave them a somewhat curious appearance *(Plates 211 & 212)*. They ran with type M3 tenders throughout *(Table 1 — Chapter 1)*.

All were scrapped between 1927 and 1929 and we have only confirmed three painted in LMS colours (Nos. 17991/4/7 — all Code C1). There may well have been more, but we do not believe any received post-1927 livery. According to many sources they were disliked by the shed and running staffs and, being non-standard, it is hardly surprising that the LMS was not excessively interested! However, they did hold the distinction, for what it was worth, of carrying the highest running numbers of any of the LMS fleet — duplicate 2XXXX series excepted, of course.

Plate 211 (Left) A broadside view of 0-8-0 No. 17997 (ex-CR No. 607) clearly displays the unequally divided wheelbase of the McIntosh 0-8-0s — livery Code C1.
Authors' Collection

Plate 212 (Right) A left-hand side view of 0-8-0 No. 17994 (ex-CR No. 604) — livery Code C1.
A. G. Ellis

Chapter 5
Glasgow and South Western Railway — Introduction and Passenger Classes

Introduction

The Glasgow and South Western Railway, often affectionately referred to simply as the 'Sou-West', formally assumed its proper title on 28th October 1850, when the main line between Glasgow and Carlisle, via Kilmarnock and Dumfries, was opened. The name of the railway had, in fact, been agreed as early as July 1846 when the Carlisle extension from the terminus of the Glasgow, Paisley, Kilmarnock and Ayr Railway was approved by Parliament. In effect, the new title was merely a renaming of the GPKAR to fit the revised geography of the enlarged system.

This 'alternative' main line between Glasgow and Carlisle (the Caledonian line — *Chapter 1* — being the original link) eventually formed the base of a sort of triangular region (with Glasgow, Carlisle and Portpatrick as the three apexes) within which the expanding G&SWR had things very much its own way in its own area. It was, of course, somewhat different where the railway came into direct contact with its 'auld' enemy, the hated 'Caley'! These two enjoyed much the same uneasy relationship with each other as did their respective English partners, the Midland and the LNWR. The difference after 1922 was that while in England, the Midland (the G&SWR's partner) had gained pre-eminence on the LMS, in Scotland, the 'other' side (i.e. the Caledonian) gained the ascendancy, at least as far as locomotive matters went — and that is our concern here.

To illustrate the extent to which the G&SWR 'drew the short straw' as it were, it is only necessary to quote the fact that ten years after the Grouping, 87.3 per cent of the Caledonian engines were still in service, 63.6 per cent of the Highland, and only 20.6 per cent of the G&SWR. As we have explained in the first chapter of this volume — and refer to again in *Chapter 7* — this was in part because of the LMS decision to standardise Caledonian components on the Scottish pre-group designs, and the G&SWR engines, for the most part, could not be modified to accept CR fittings as easily as could many of the ex-HR types. But this was not the whole story. The G&SWR had more than 500 engines at the Grouping — some three times the total HR fleet — yet their destruction was rapid and wholesale, and cannot fully be comprehended without a brief explanation of the other historic causes, quite aside from their incompatibility with Caledonian designs. We therefore feel it necessary to give a little more background detail before going on to the detailed fleet analysis.

The G&SWR was a railway full of character and, indeed, as other writers have averred, full of 'characters' as well! Many of these were in the locomotive department and for most of its existence, the Company looked to its own resources — or, more specifically, to men who had been trained by or worked at Kilmarnock — to provide it with adequate motive power. This they did to a remarkably successful degree until 1912, when the continuity was abruptly broken after the retirement of James Manson as Locomotive Superintendent and the appointment of the first 'outsider', Peter Drummond, to succeed him.

Drummond, who died in office in 1918, was succeeded by another outsider, Robert Whitelegg and, it has to be said, that neither of these engineers did the G&SWR many favours. The situation was not one of unrelieved gloom, but neither was it particularly memorable, except for its adverse effects. It is no part of our task to go into great detail, but just as with the LMS as a whole one cannot understand the locomotive story without considering personalities (and conflicts — *see Volume One*), so too with the G&SWR, the unhappy fate of its engines in LMS days was in part a legacy of events before 1923.

Locomotive affairs on the G&SWR had proceeded in a reasonably orderly fashion from the days of the famous Patrick Stirling through the development of his younger brother James, who was in turn followed by Hugh Smellie and James Manson. All were fine engineers in their own ways, and all were 'Sou-West' men in inclination or training. Each built on the foundations of his predecessor by development and/or rebuilding of earlier types, and their products were 'in tune' with the character and needs of the Company. With but a few exceptions, however, G&SWR locomotives were a little on the small size by 1912.

Bearing in mind that Patrick Stirling's appointment dated from 1853, there had been nearly sixty years of continuous evolution when Peter Drummond came on the scene and began to upset everything. He changed the whole character of the locomotive scene in many ways. He scorned the small engines of his predecessors and followed the precepts of his brother Dugald (on the LSWR) for design ideas. He moved the driver to the opposite side of the cab and, at least for a few years, built new and big engines which by any objective standards were not very good. Towards the end of his short period in office, he was beginning to be more understood, his later engines were better, and he may have been about to 'turn the corner' as it were, when he died — we shall never know.

Whitelegg, in succession to Drummond, had an even more unenviable job. The railway was, like many, run-down after World War I, quite apart from the Drummond hiatus, and Whitelegg realised the need to get to grips with both problems. He consulted staff more and tried to find a blend of the best of Manson and Drummond. He realised that much of the stock was overdue for replacement but, for financial reasons, had to resort to extensive rebuilding rather than total renewal. A new range of boilers was planned, some based on Drummond's new engines, and others to his own design.

Three of these (the X1, X2 and X3 types) had appeared before the Grouping and were fitted to many of the earlier Manson designs.

All this was most laudable and deserved a better outcome, but Whitelegg's rebuilds were frequently inferior to the original machines (and disliked accordingly) and he ran out of time as an independent locomotive chief when the LMS was formed. It was all very sad. The ten years of backward progress after Manson's retirement were, therefore, as much a cause of the 'slaughter' of the G&SWR during the early LMS period as any incompatibility with CR types.

The natural consequence of all this was that the new Midland-dominated LMS management felt that the G&SWR fleet was more a liability than an asset, and acted accordingly. It was less than fair to the excellent work of the earlier engineers like James Stirling, Hugh Smellie and James Manson, but was, in the circumstances quite inevitable.

Not the least of the LMS problems was the almost total lack of standardisation on the G&SWR section. Had Drummond been less draconian or Whitelegg more successful, or had both of them continued to develop the fine work of Manson, then the LMS would probably have had a reasonably harmonious group of 500 or more engines to inherit which may have fared rather better. As it was, they represented possibly the most bewildering array of types (relative to fleet size) inherited by the LMS, and we will need to devote proportionally more space in order to analyse even their short LMS lifespan — rarely much more than ten years — just as we have found it necessary to write this rather more extended introduction to our analysis.

For one thing, there was no dominant 'house' style as, for example, on the LNWR, Caledonian or LYR. Neither was there real standardisation of detail as on the Midland. We therefore feel that since our principal aim is to concentrate on outward appearance, a brief resumé of the visual characteristics, still visible in 1923, might be helpful.

a) The Stirling engines — By 1923, one or two engines only were left from the Patrick Stirling period and can be disregarded. However, his younger brother, James, had followed very closely his brother's ideas, and quite a number of engines dating from his period of office (1886-78) were acquired by the LMS, little changed, visually, from their original form. A characteristic example is shown in *Plate 213*.

Plate 213 0-4-4T No. 731 shows the characteristic James Stirling features (rounded cab, deep footplate angle) which several G&SWR engines carried through to the LMS. No. 731 was allocated the number LMS 15244, but never carried it.
W. Stubbs Collection

b) Hugh Smellie — Smellie was James Stirling's Works Manager in the early days until 1870, came back in 1878 to the G&SWR from the Maryport and Carlisle Railway *(see Volume Two, Chapter 12)* and soon reverted to G&SWR ways! His engines were a natural development from those of Stirling (domeless boilers, round-topped cabs, etc.), gradually getting larger and generally more elegant in outline. Some of them were distinctly beautiful and a few reached the LMS, little modified. *Plate 214* shows a good example. Smellie was a fine engineer, fully deserving his elevation to the larger Caledonian job *(see Chapter 1)* but he died only a few months later, so his G&SWR work could be said to be his last real achievement.

Plate 214 Hugh Smellie also favoured a rounded cab shape and domeless boiler, but gave his engines altogether more refined proportions. This is 153 class 4-4-0 No. 14156 (ex-G&SWR No. 464) in red LMS livery, Code A1.
M. J. Robertson Collection

c) James Manson — Manson was the third consecutive 'locally trained' chief when, in 1890, he assumed the mantle of G&SWR locomotive affairs. Like Smellie, he had 'cut his teeth' as Locomotive Superintendent on a lesser concern (The Great North of Scotland Railway) before returning to Kilmarnock where he had trained. In developing Smellie's work, he gradually introduced changed visual lines, particularly in terms of cab shape and domed boilers, and his engines were almost always 'good lookers'. At his best, their visual lines could rival those of Johnson on the Midland Railway — and we can think of no higher compliment. Many Manson engines came, substantially unaltered, to the LMS and *Plate 215* is typical of his approach.

Plate 215 This view of LMS No. 14177 (ex-G&SWR No. 398) shows one of the handsome Manson 8 class 4-4-0s, virtually 'as built'. The changed cab shape, built-up chimney and the domed boiler are, perhaps, the most characteristic features of Manson's designs. The LMS livery code is A1.
Photomatic

d) Peter Drummond — Drummond changed everything — or seemed to. His intention was to build 'big' engines for the G&SWR and they were built like battleships. In railway terms they were about as useful too — but perhaps we are unkind! He followed his brother's LSWR precepts very closely, and the visual lines of his engines were entirely in the Drummond idiom — more familiar on the Caledonian, LSWR and, to some extent, the Highland. They did not lack in proportion *(Plate 216)* but were completely alien in outward form to G&SWR tradition. Perhaps if they had worked better, it would have helped.

Plate 216 The massive Drummond 'look' is well-caught by 4-4-0 No. 14515 (ex-G&SWR No. 336) in rather dirty LMS lined black with 12in. figures, Code B3, and photographed some time after the engine had been superheated, probably circa 1930.
Authors' Collection

e) Robert Whitelegg — Whitelegg was perhaps one of the unluckiest of engineers. He had already been frustrated on the LT&SR *(see Volume Four)* and this story was to be repeated on the G&SWR. Visually, he added a fifth element to the G&SWR fleet. His one new design — the 4-6-4T, *page 136* — was a celebrated type, but his rebuilding of the earlier engines was visually characterised, more often than not, by a somewhat higher-pitched boiler (of new type), and a new cab style with extended roof and almost 'Gothic' characteristics where the upright rear stanchions met the roof. *Plate 217* shows this very well, and the change in visual character compared with the original Manson/Stirling/Smellie product was quite marked and quite different from all previous approaches. With his rebuilds of the 4-4-0s and 0-6-0s, Whitelegg normally modified the tenders to carry coal rails — a conceivably prophetic move! These coal rail tenders generally, but not exclusively, remained with the rebuilds throughout.

Plate 217 Whitelegg built very few new engines for the G&SWR, but his cab shape and circular smokebox handrail were distinctive on most rebuilds. No. 14130 (ex-G&SWR No. 468), LMS livery Code A1, shows these features on a rebuilt example of the Smellie-designed 'Wee Bogie', originally of the type shown in *Plate 219*. At a later date, most of the circular smokebox handrails were replaced — see, for example, No. 14120 *(Plate 222)*.
Authors' Collection

In our analyses of ex-G&SWR engines which follow, we shall refer to locomotives by their original designer and classification. The latter generally reflected the running number of the first member of the class to be built. The whole fleet was renumbered in 1919 — which numbers we quote in captions, etc. — but most classes derived their nomenclature from the pre-1919 number series, and this accepted custom will be followed here.

Additionally, we should, by way of preliminaries, point out that the G&SWR was essentially a tender engine railway with but a small proportion of tank types. There was a time, in fact, when it was almost an obscenity even to mention tank engines at Kilmarnock! Thus, we have divided our survey merely into passenger and freight rather than adopt the four-fold sub-division used with the other larger LMS constituents.

For all its lack of relevance to the broader LMS scene, the G&SWR locomotive contribution was by no means the least interesting. For our part, we find it one of the most fascinating stories we have researched, and if our affection for its products spills over into our writing about them, we do not propose to apologise!

Passenger Tender Classes

2-4-0 Classes — all types

Depending on how one defines it, three or four engines of the 2-4-0 wheel arrangement reached the LMS. One, G&SWR No. 726, was scrapped in 1923 before the LMS number series was issued, and its sister engine, No. 727, was allocated LMS number 14000 — the first number in the Northern Division series. Both were members of the 75 class, designed by James Stirling in 1870 and, regretfully, we have been unable to find a picture of No. 727. It did not receive its LMS number.

The other two 2-4-0s (LMS allocated Nos. 14001-2) were members of Hugh Smellie's 157 class, introduced in 1879-80 as a sort of retrospective revival of the 2-4-0 style after Stirling had introduced the 4-4-0 type. However, Stranraer could not turn a 4-4-0 at that time, so the new 2-4-0s were used, inter alia, on boat trains. Twelve were built, but the two LMS survivors did not receive their new numbers in 1923, both being withdrawn the same year. We illustrate one of them in G&SWR colours in *Plate 218*.

Plate 218 Hugh Smellie's 157 class 2-4-0 No. 723 is shown just before scrapping in 1923. It should have become LMS No. 14002, but never did.
Authors' Collection

4-4-0 Classes

Apart from the Manson 4-6-0s *(below)* the G&SWR placed almost total reliance on the 4-4-0 type for most of its longer-distance passenger working. The LMS inherited 181 examples from no fewer than nine nominal classes and a considerably greater number of visual variants. We deal with them, as usual, in ascending LMS numerical order, using their original designer and class designation as points of reference.

Smellie 119 Class (LMS Nos. 14116-37; Power Class 1/2, later 1P/2P)
These were not the oldest ex-G&SWR 4-4-0s but, possibly by virtue of their size, went to the head of the LMS number lists. Known as the 'Wee Bogies', they were highly regarded engines, originally designed for use on the Greenock services when introduced in 1882. They were very similar to the same designer's 2-4-0 in original form, and the LMS received eight examples substantially as built *(Plate 219)*. These were Nos. 14116/22/4-5/7/32/4-5. At any one time, two of these could be seen carrying domed boilers, introduced by Manson in 1909. We have confirmed Nos. 14122/4/7/35 carrying these boilers at some time during the LMS period as boilers were changed around, sometimes with wingplates *(Plates 220 & 221)*.

Plate 219 'Wee Bogie' No. 14132 (ex-G&SWR No. 715) depicts the type virtually 'as built' with domeless boiler and round-cornered cab, but without smokebox wingplates. The LMS livery code is B3 and the tender has Whitelegg pattern coal rails.

M. J. Robertson Collection

Plates 220 & 221 These two views show examples of the Smellie 119 class carrying small-domed boilers both with and without wingplates. The engines are Nos. 14124 (ex-G&SWR No. 709) and 14122 (ex-G&SWR No. 707), both in LMS red livery, Code A1.

Authors' Collection

Twenty four 'Wee Bogies' were built and Drummond began their withdrawal, but only two had gone when World War I caused a stop to further scrapping and the twenty two survivors all reached the LMS. By this time, those fourteen members of the class not already considered above had been rebuilt circa 1921/2 with the Whitelegg cab and X3 boiler, completely transforming their appearance *(Plate 222)*. They were not quite as awful as some Whitelegg rebuilds but, according to David L. Smith, they lacked the 'sparkle' of the performance of the original version. They were, none the less, put in Power Class 2/2P by the LMS, rather than the 1/1P of the originals, having higher boiler pressure and, in consequence, higher nominal tractive effort. Scrapping took place from 1925-34, the last to go being Nos. 14118/20 (both rebuilds). No. 14125 (unrebuilt ex-G&SWR No. 710) never received its LMS number.

Plate 222 Whitelegg's rebuild of a 'Wee Bogie' is represented by No. 14120 (ex-G&SWR No. 704) in LMS pre-1928 crimson, Code A1.

Photomatic

In LMS livery terms, the 119 class was straightforward. The vast majority received the full pre-1928 red livery, Code A1, and of the few which received post-1927 liveries, lined black with 12in. figures (Code B3 or Code B6) seems to have been favoured. These styles are given in *Plates 219 & 220*. It seems likely that the majority were withdrawn in their pre-1928 colours. As usual, it is difficult to be certain about whether the lined black liveries had plain insignia (Code B6) or red shaded (Code B3).

Livery Samples

Code A1 (smaller boiler)   14116/22§/4§/7§/32/4-5§
Code A1 (rebuilt)   14117-20/3/6/8-30/3
Code B3 (small boiler)   14312
Code B3 (rebuilt)   14120/36

Notes: Letter spacing, post-1927, at 53in. centres
  § Small boiler with dome

Smellie 153 class (LMS Nos. 14138-56; Power Class 1/2, later 1P/2P)

The 'Wee Bogies' had but 6ft. 1½in. driving wheels, and Smellie's next design of 4-4-0, the 153 class, was essentially the same type but carrying 6ft. 9½in. wheels of 'express' diameter. Introduced in 1886, twenty were built and all but one reached the LMS. In many ways they were probably the best 4-4-0s built for the G&SWR and, in original form, were beautiful and graceful-looking engines *(Plate 223)*. They took over the 'crack' workings from the Stirling 4-4-0s *(below)*.

Plate 223 The graceful lines of Hugh Smellie's 153 class 4-4-0s as originally built are well displayed by No. 14141 (ex-CR No. 459) in rather grubby LMS red livery (Code A1), at Kilmarnock on 24th May 1927.

A. G. Ellis

Like the smaller-wheeled 4-4-0s, some three of the 6ft. 9½in. engines were fitted with a Manson domed version of the original boiler. In 1922, Whitelegg began to rebuild them with the X2 boiler to the form shown in *Plate 225*, and by early LMS days, eleven had been so treated, including two of the Manson domed versions. Some of this rebuilding actually took place during 1923, but after this time, the process stopped, and the following engines ran on in original form until scrapping — Nos. 14138/41/8/51-4/6. Of these, No. 14153 was the residual sole survivor with domed Manson boiler — *Plate 224*.

Plate 224 The only long-lasting survivor of the 153 class with the small Manson-domed boiler was No. 14153 (ex-G&SWR No. 461), seen here at Glasgow (St. Enoch) circa 1926/7 in LMS red, Code A1.

Authors' Collection

Plate 225 This is a Whitelegg type 153 class rebuild actually carried out by the LMS. No. 14143 was G&SWR No. 466 and is pictured in lined black livery, Code B3.

Photomatic

LMS liveries were as for the 'Wee Bogies', namely pre-1928 red (Code A1) and post-1927 lined black with 12in. figures (Code B3 or Code B6), entering the usual caveat about insignia shading (black or countershaded red). Finally, Nos. 14138/48/51/2/4 never received LMS numbers.

Livery Samples

Code A1 (smaller boiler)	14141/53§/5
Code A1 (rebuilt)	14142-7
Code B3 (rebuilt)	14139-40/3-4

Notes: Letter spacing, post-1927, at 53in. centres
  § small boiler with dome

Manson 8 and 240 classes (LMS Nos. 14157-202/244-70; Power Class 1/2, later 1P/2P)

This group of 6ft. 9½in. engines was probably the most confusing of all the ex-G&SWR 4-4-0 series. It was made worse by the somewhat chaotic way in which the LMS renumbered them, so we have put them all together for simplicity of treatment.

Essentially, the series consisted (as built) of 57 engines to Manson's 8 class, introduced in 1892 — a supremely elegant design of considerable merit, of which the majority passed through, unmodified, to the LMS in the form represented in *Plate 226*. These 57 engines were allocated LMS Nos. 14157-202/44-5/9-53/66-9. They were followed in 1904 by fifteen of the 240 class, an identical 'engine' portion, but fitted with a larger boiler and higher cab. These were given LMS Nos. 14246-8/54-65 and an example is given in *Plate 227*. Two of the engines were given special eight wheel tenders to allow the midday Anglo-Scottish trains to make longer runs between water stops. At first attached to 8 class engines, they ended up attached to 240 class examples — *Plate 228*. Finally, during 1910-12, five of the 8 class were rebuilt by Manson to the 240 class type and this, in effect, concluded the Manson part of the story with fifty two 8 class (as built) and twenty 240 class (fifteen original plus five rebuilt). The five 240 class rebuilds were LMS Nos. 14168/266-9.

1802.

14261

1804.

Plate 226 (Above Left) The original Manson 8 class 4-4-0s had something of a Johnson look to them — even the built-up chimney was reminiscent of Midland styling. This example is No. 14182 (ex-G&SWR No. 410) in a somewhat dirty state, LMS livery Code A1.

Authors' Collection

Plate 227 (Below Left) No. 14261 (ex-G&SWR No. 389) shows one of Manson's 240 class 4-4-0s built in this form and never subsequently rebuilt. Note the many detail similarities with the 8 class *(Plate 226)*, even though the boiler is self-evidently larger and pitched higher. The LMS livery code is A1.

Authors' Collection

Plate 228 (Above) LMS No. 14248 (ex-G&SWR No. 381) was a 240 class 4-4-0 built as such but equipped with Manson's unusual eight wheel tender *(see text)*. The leading four wheels formed a bogie but the two rear axles were fixed. The engine is in LMS red, Code A1.

Stephen Collection, courtesy NRM

Plate 229 (Below) LMS No. 14169 (ex-G&SWR No. 406), livery Code A1, shows an 8 class 4-4-0 as rebuilt by Whitelegg. Although not as good as the originals, these engines still retained a characteristic neatness of outline.

Stephen Collection, courtesy NRM

After the Drummond era, Whitelegg started to rebuild further examples of both Manson types to the form shown in *Plate 229*. Apart from their cabs, they showed considerable visual affinity to the original 240 class but, as traffic machines, they were disastrous compared with the genuine Manson versions. Most of the Whitelegg rebuilds, fifteen in all, came from unmodified 8 class engines (LMS Nos. 14158/60/2/9-71/3/5/83/5-6/92/200-1/44/52), but two of the 240 class were also rebuilt in 1920 and became LMS Nos. 14246/60. One or two of the Whitelegg rebuilds had a slightly different cab roof *(Plate 230)* and one of them received one of the eight wheel tenders *(Plate 231)*. Finally, the last member of the group (LMS No. 14270 — *Plate 232*) was built new in 1921 to the Whitelegg variant. Thus, by LMS days, the 73 members of this 6ft. 9in. series were divided as follows:

Class 8 (as built)	36 examples
Class 240	18 examples (5 ex-8 class)
Whitelegg rebuilds	18 examples (16 ex-8 class, 2 ex-240 class)
Whitelegg built new	1 example

The LMS classified the larger boilered engines (240 class and Whitelegg type) as Power Class 2/2P, the original engines as 1/1P.

Plate 230 (Left) A few Whitelegg rebuilds sported a slightly different cab shape, with the main visible variation where the upright stanchion met the cab roof. This is shown on No. 14260 (ex-G&SWR No. 388), rebuilt from a 240 class 4-4-0 and seen at Hurlford in LMS red (Code A1) on 22nd May 1928.

A. G. Ellis

Plate 231 (Below) The combination of Whitelegg cab and boiler, fully circular smokebox handrail and Manson 8 wheel tender was, we believe, unique to No. 14246 (ex-G&SWR No. 379), a 240 class rebuild seen here at Hurlford in May 1927 in red livery, Code A1.

A. G. Ellis

Plate 232 (Below) Locomotive No. 14270 (ex-G&SWR No. 485) was somewhat unique in having been built new by Whitelegg to his rebuilt form of the Manson 6ft. 9½in. engines. This broadside view gives a good impression of the first LMS livery, Code A1.

BR (LMR)

In terms of LMS livery, as was customary on the G&SWR section, there was considerable consistency. Before 1928, all repaints were red with 18in. figures (Code A1) and the post-1927 lined black liveries we have verified *(Plates 233 to 235)* suggest 12in. figures for the Whitelegg rebuilds and 10in. figures for both of the Manson versions. All varieties were extinct by the end of 1933, and most were probably withdrawn in pre-1928 livery. Several did not receive LMS numbers before withdrawal but we cannot offer 100 per cent confirmed details. Nevertheless, we can confirm that the following engines never received LMS numbers: Nos. 14162/4/6/74/81/7/90-1/3/249/54/66. There may have been a few more if readers could help.

Plates 233 to 235 These three similarly-posed views enable easy comparison to be made between the three forms of Manson 6ft. 9½in. 4-4-0s — all in post-1927 livery. No. 14172 (ex-G&SWR No. 421) represents the original 8 class (livery Code B2), No. 14267 (ex-G&SWR No. 378) represents the 240 class (livery Code B2), in this case rebuilt from an 8 class engine, while No. 14183 (ex-G&SWR No. 426) is a Whitelegg 8 class rebuild, livery Code B3.
Photomatic and Authors' Collection

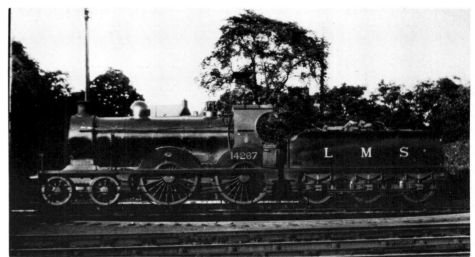

Livery Samples

8 class (as built)
Code A1   14161/5/72/6-80/2/4/95-6/9/202/45/51/3
Code B2   14172/94-6

240 class (as built/rebuilt)
Code A1   14168/247-8/55-9/61-5
Code B2   14248§/58/67

Whitelegg type (all versions)
Code A1   14160/9/71/3/5/85×/92/200×/1×/44/6§/60×
Code B3   14160/70/3/83/246§

Notes: Letter spacing, post-1927, at 53in. centres
   § *eight wheel tender*
   × *Modified Whitelegg cab, as Plate 230*

Manson 336 class (LMS Nos. 14203-27; Power Class 2, later 2P)
This group of highly effective engines was introduced in 1895 as Manson's version of a 6ft. 1½in. 4-4-0 for the Greenock services, in succession to the 'Wee Bogies'. Twenty five in all were built and all came to the LMS, the majority unmodified. They were an adaptation of the 8 class with smaller wheels and higher boiler pressure, and were considered LMS Power Class 2 at all times, whether original or rebuilt *(below)*. They shared many visual similarities with the 8 class, and were equally handsome engines *(Plates 236 & 237)*.

Plates 236 (Below) & 237 (Top Right) These two views show Manson 336 class 4-4-0s, as built. No. 14212 (ex-G&SWR No. 359) retains its Manson chimney and smokebox door, while No. 14209 (ex-G&SWR No. 356) has a replacement Whitelegg chimney and smokebox door. Both engines are in LMS red, Code A1.

Authors' Collection and A. G. Ellis

As usual, Whitelegg made an attempt at rebuilding some of them with X1 boilers and six had been thus treated by the time of the Grouping. These became LMS Nos. 14204/8/11/13-4/23. Like most Whitelegg rebuilds, they retained a neat outline *(Plate 238)* but yet again their performance was not improved, although being so good originally, the rebuilds were not quite so dreadful as some of Whitelegg's efforts.

The whole series was scrapped by the end of 1932 and we have no record of the following ever receiving LMS numbers: Nos. 14205/7/10/8/21/5. However, only Nos. 14205/10 are positively confirmed as not so doing.

The livery of these engines in LMS days was quite straightforward. We believe that all which received LMS numbers were painted red in the pre-1928 period (Code A1), and have confirmed most of them *(below)*. Most were scrapped in this livery but a few received lined black *(Plates 239 & 240)* with variations in numeral size to be observed. As usual, it is impossible to be certain of insignia colour, but we think red shading to be more probable and have listed them accordingly.

Plate 238 No. 14214 (ex-G&SWR No. 361) is a Whitelegg rebuild of a Manson 336 class engine. It is seen at Glasgow (St. Enoch) in red livery (Code A1) circa 1926.

Authors' Collection

Plates 239 & 240 No. 14211 (ex-G&SWR No. 358) was another Whitelegg 336 class rebuild and is pictured carrying two varieties of the post-1927 livery, Code B2 (with 40in. letter spacing) and Code B4 (with normal letter spacing). Note the removal of smokebox numberplate sometime between the taking of these two pictures.
Authors' Collection

Livery Samples

Original Condition
Code A1   14203/6/9/12/5-7/9/22/4/6-7
Code B3   14212

Rebuilt Condition
Code A1   14208/14
Code B2   14211 (with 40in. spacing of 'LMS' — Plate 239)
Code B3   14204
Code B4   14211

Note: Letter spacing, post-1927, at 53in. centres, except where noted

Stirling 6 class (LMS Nos. 14228-43; Power Class 1, later 1P)
The oldest G&SWR 4-4-0s were these engines to James Stirling's design of 1873. At that time, a large-wheeled (7ft. 1½in.) 4-4-0 was a distinctly adventurous concept for the G&SWR but the 22 examples built gave yeoman service on the main line expresses until the turn of the century. By this time they were almost life-expired, and six had already been scrapped. However, during 1899-1901 Manson rebuilt the sixteen survivors with new cabs and Stirling type boilers and in this form, all sixteen came through to the LMS *(Plates 241 & 242)*. Two of the sixteen boilers had domes, and these two were exchanged around various members of the class at shopping *(Plate 243)*.

We can give no reason, except perhaps for wheel diameter, why these engines were given such high LMS numbers in 1923, nor why they were inserted, along with the Manson 336 class, in the middle of the Manson 6ft. 9½in. series. There was some logic with the 336 class, being LMS Power Class 2, but the Stirling engines were Power Class 1. Perhaps readers can help.

The 6 class was an outstanding success. Built, in effect, to reflect the opening of the Midland/G&SWR route from London to Scotland via the Settle and Carlisle line, they gave fifty years of service before the Grouping and most of them lasted several years beyond as well. They were probably James Stirling's most outstanding design, and even though they lost their characteristic 'rounded' cabs at rebuilding, much of the original remained. The last examples (nos. 14234/6) lasted until 1930, but several did not receive LMS numbers. We believe these to have been Nos. 14228-31/8/43, all but No. 14238 being positively confirmed.

Plates 241 & 242 Opposite side views of Stirling 6 class 4-4-0 No. 14232 (ex-G&SWR No. 473) as received by the LMS with domeless boiler and 'old type' tender — *see text*. The left-hand side view was taken at Hurlford in May 1927, the year of the engine's withdrawal. The livery code is A1.

A. G. Ellis and Authors' Collection

Plate 243 This view of No. 14234 (ex-G&SWR No. 475) shows the combination of domed boiler and replacement tender on a Stirling 6 class engine. The livery is red, Code A1.

BR (LMR)

The LMS painted at least eight of them in red livery (Code A1) and we do not believe any received the post-1927 style. However, during LMS days, the majority received newer tenders of Smellie's design from withdrawn engines *(Plates 243 & 244)*. These included Nos. 14233-6/40-2. We summarise the variations below.

Livery Samples — all Code A1

Domeless boiler, old tenders	14232/7
Domeless boiler, new tenders	14235-6/41-2
Domed boiler, new tenders	14233-4

Plate 244 This picture shows the most common LMS version of the 6 class — domeless boiler with replacement tender. No. 14236 was ex-G&SWR No. 477 and is seen in red livery, Code A1.

Stephen Collection, courtesy NRM

Manson 18 class (LMS Nos. 14366-78; Power Class 2, later 2P)

The Manson 18 class introduced in 1907, was effectively, a development of the 240 class *(above)* using the same 'engine' components but a revised and redesigned boiler and firebox, the latter being long and shallow. In practice, therefore, they represented the end of a line of evolution which began with Smellie's 153 class and, as usual, Manson produced another handsome design — *Plates 245 & 246*. The final example (LMS No. 14378, ex-G&SWR No. 345) came out a little later than the main batch (in 1912) and was Manson's last new engine before he retired.

Plates 245 & 246 Opposite side views of Manson 18 class 4-4-0s Nos. 14366 (ex-G&SWR No. 337) and 14375 (ex-G&SWR No. 347) both in red livery, Code A1, give a clear impression of Manson's final 4-4-0 type.
A. G. Ellis and BR (LMR)

In fact, the engines, although satisfactory, proved no real improvement on the earlier 240 class, and gained a reputation for rough riding. Perhaps it was for this reason that they lasted no longer than the rest of the ex-G&SWR 4-4-0s, being withdrawn between 1925 and 1932. One only is positively confirmed as not having received its LMS number (No. 14373).

An interesting detail point was the new design of flat-sided tender — somewhat reminiscent of Midland practice at the time but not quite the same. All members had this tender, although, for a while in G&SWR days, two examples were temporarily paired with the eight wheel tenders built earlier — *see page 115*. This pairing did not last until the LMS period.

LMS liveries were straightforward with one noticeable, and well photographed, exception. Before 1928, full crimson lake, Code A1, was the 'norm', but No. 14377 received the full freight livery (plain black — Code C1) for some reason. We illustrated this in *Plate 202 of Volume One*. After the livery change, a few were turned out in lined black, and the one really clear source *(Plate 247)* undoubtedly shows countershaded insignia, which is why we tend to believe that most ex-G&SWR engines which received the LMS intermediate livery were probably given the red shaded characters. Most received 12in. figures rather than the 10in. style shown.

Plate 247 No. 14374 (ex-G&SWR No. 346) is shown in lined black livery, Code B2, clearly bearing countershaded insignia. The 12in. numerals were somewhat more common on this type with this style of painting.

Photomatic

Livery Samples

Code A1   14366-9/72/5
Code C1   14377
Code B2   14374
Code B3   14366/9/70/6

Note: Letter spacing, post-1927, at 53in. centres

Manson/Whitelegg experimental 4-4-0 (LMS No. 14509; Power Class 3, later 3P)
In 1897, Manson built a 4-4-0 virtually identical with his 8 class, but having four cylinders, the first such machine in Britain. It was given G&SWR No. 11 and was sometimes referred to as the 11 class. It was generally regarded as less economical than the 2-cylinder engines and in 1915 was rebuilt with the larger 240 class boiler. In 1922, Whitelegg virtually renewed the whole engine with a large Drummond superheated boiler and his own characteristic cab.

This massively proportioned rebuild was named *Lord Glenarthur* and came to the LMS, virtually brand new, in this much modified form. It did some quite good work on the Ayr services but was scrapped in 1934. Its LMS history is well summarised in *Plates 248 to 250*.

Plates 248 (Left), 249 (Above) & 250 (Right) *Lord Glenarthur* (ex-G&SWR No. 394) is shown as built with full circular smokebox door handle, as later modified and as finally running in lined black livery. The livery codes are A1, A1 and B3 respectively.

A. G. Ellis and Authors' Collection

Drummond 131 and 137 classes (LMS Nos. 14510-21; Power Class 3, later 3P)
Drummond's first 4-4-0 in 1913 for the G&SWR must have come as a great shock to the system for it was the heaviest of the type yet seen in Britain and designed with small diameter (6ft.) driving wheels, ostensibly to replace the dainty 'Wee Bogies' — *see page 112*. The first six of the new type (the 131 class) were saturated engines and very much inspired by Dugald Drummond's LSWR design of similar proportions. Numbered 14510-5 by the LMS, they came to the new company in saturated form, but in 1923 (No. 14510) 1926 (Nos. 14513-5) and 1931 (No. 14512), five were fitted with superheated boilers of the 137 class type. Thus, No. 14511 *(Plate 251)* was the only engine of the original six to remain saturated throughout the LMS period. It could be identified by its (slightly) shorter smokebox.

Plate 251 No. 14511 (ex-G&SWR No. 332) was the only Drummond 4-4-0 of the G&SWR never to be superheated. It is pictured at Glasgow (St. Enoch) circa 1930 in very dirty lined black livery, Code B7.

M. J. Robertson Collection

The 131 class were ponderous and sluggish machines and, in 1914/15, Drummond at last saw the virtues of full superheating and built six more engines, the 137 class, with Schmidt superheaters (LMS Nos. 14516-21). The weight went up again to establish yet another British record for a 4-4-0 until Gresley's LNER Class D49 appeared in 1927. The 137 class were much better than the original six and began to carry out excellent work on the boat trains. They were swift and economical, but many G&SWR men, who had good cause to suspect Drummond's engines, based on their 0-6-0 experience — *see Chapter 6* — had no time for them, and they gained only a patchy regard. *Plate 252* shows one of this series.

Plate 252 LMS No. 14517 (ex-G&SWR No. 326) was built superheated — note the larger smokebox compared with No. 14511 *(Plate 251)*. The livery is lined black and the insignia are clearly black shaded with 'highlight' lines to the figures — Code B7.

A. G. Ellis

Plate 253 This 131 class locomotive, No. 14515 (ex-G&SWR No. 336) was superheated in 1926 and we guess this picture was taken soon afterwards — livery Code A1. A later view of this engine is given in *Plate 216*.
A. G. Ellis

However, in later years when the 131 class superheat rebuilds were also at work, the Drummond 4-4-0s were by no means outclassed by some of the new LMS types and lasted well into the mid-1930s, escaping the 1923-33 slaughter. Thus, in retrospect, Drummond's reputation can be said to have been partially retrieved by these engines. One suspects, admittedly with the benefit of hindsight, that it was the manner of the changes as much as the nature of them which caused Drummond his problems on the G&SWR, and an interesting parallel can be drawn with Edward Thompson's take-over from Gresley on the LNER in 1941.

During LMS days this group of twelve engines lasted long enough to receive both liveries, although we have failed to confirm many with the pre-1928 red livery, Code A1 — *Plate 253*. We feel sure that most, if not all of them must have received the style.

After the 1927/8 change, smokebox numberplates were usually removed and lined black was adopted — frequently obscured under layers of grime — and 14in. Midland figures were by far the most common *(Plate 252)*. However, one or two examples received 12in. figures *(Plate 254)*. The last of the engines (No. 14513) was withdrawn at the end of 1937, and we believe all carried the post-1927 lined black livery mostly with red-shaded characters, but not exclusively — *see Plate 252*.

Plate 254 Superheated 131 class No. 14514 (ex-G&SWR No. 335) bearing the somewhat less common (for this class) lined black livery with 12in. figures, Code B3.

A. G. Ellis

Livery Samples

Code A1	14510/3/5/9
Code B3	14514-5
Code B4	14510/3/6/8/20-1
Code B7	14511-2/7

Notes: a) All engines in superheated condition except No. 14511
b) Letter spacing, post-1927, at 53in. centres or slightly larger
c) Lined black codes are the 'most' probable — see text

4-6-0 Classes — all types

Manson 381 class (LMS Nos. 14656-72; Power Class 3, later 3P)

This group of handsome and elegant engines came into service in two separated batches. The first ten (LMS Nos. 14656-65) were built in 1903 and the last seven (LMS Nos. 14666-72) in 1910/11. They were designed to eliminate double-heading on the main line, and to a fair extent this happened, but loads, especially in summer, continued to grow so the 4-6-0s sometimes still needed a pilot engine. Nevertheless they were fine machines after some initial teething troubles and did good work in the Manson period.

Built without superheaters, they were never subsequently given this fitting but Whitelegg (inevitably?) started to experiment with them — to no great avail. In fact, and not for the first time, his ministrations proved to be retrograde. Extended smokeboxes were fitted which adversely affected the steaming, and two new Whitelegg boilers were made, along with new cabs for the two engines concerned, but nothing was achieved and much was destroyed. In consequence, by the post-1918 period, the Manson 4-6-0s were ill-regarded by outside observers. It need not have been so. A more drastic rebuild along the lines of the 128 class *(below)* would have produced a superb engine.

Turning now to the visual characteristics, the first ten engines came out with distinctive bogie tenders *(Plate 255)* as did four of the final seven. The last three of these seven, however, reverted to a six wheel tender of more conventional type, and all seven engines of this final batch had modified cabside sheets with a reduced height 'cut-away'. Both these features can be seen in *Plate 256*. Most of the earlier engines carried safety rails to compensate for the low cabside *(Plate 257)* but one of them was modified *(Plate 258)* and also exchanged its bogie tender with No. 14672 for a six-wheeler. All these examples illustrated show the extended smokebox of the original boilers as they came to the LMS, but the two engines with new Whitelegg boilers (Nos. 14671-2) were also given new cabs *(Plates 259 & 260)*. In consequence, there were a number of possible permutations to be observed. In LMS days, the most common single variant was extended smokebox, low cab-side with safety rails, and bogie tender, as in *Plate 255*. One or two seem to have received Whitelegg type smokebox handrails *(Plate 261)*.

Plate 255 Manson 381 class 4-6-0 No. 14658 (ex-G&SWR No. 497) of the first series, in the condition 'as received' by the LMS — livery Code A1. It is recorded that, in 1926, this engine took a 100 wagon coal train over Ais Gill to Skipton, assisted by a Class 2P 4-4-0!

A. G. Ellis

Plate 256 (Right) The second series of Manson 4-6-0s had a revised cabside shape — shown here on No. 14670 (ex-G&SWR No. 509), one of three fitted also with six wheel tenders. The livery code is A1.

Authors' Collection

Plate 257 (Left) No. 14664 (ex-G&SWR No. 503) in lined black livery, Code B7, clearly shows the safety rails to the cabsides of the first series of Manson 4-6-0s.

Photomatic

Plate 258 (Below) No. 14659 (ex-G&SWR No. 498) exchanged tenders with No. 14672 *(Plate 260)* and also had its cabside modified. It is seen here in LMS lined black livery, Code B4.

Photomatic

Plate 259 No. 14671 (ex-G&SWR No. 510) in red livery, Code A1, was one of two Manson 4-6-0s fitted with Whitelegg boiler and cab. It was also one of three given 6 wheel tenders.
M. J. Robertson Collection

Plate 260 The other Whitelegg rebuild was No. 14672 (ex-G&SWR No. 510) pictured with the bogie tender exchanged with No. 14659 *(Plate 258)*. The livery is probably Code B7, but we cannot be positive.
Authors' Collection

Plate 261 No. 14657 (ex-G&SWR No. 496) was a rare (unique?) example of a Manson 4-6-0 with Whitelegg smokebox door and circular handrail — livery Code A1.
A. G. Ellis

Liveries were straightforward. We believe that all were painted pre-1928 crimson lake (Code A1) and the later survivors were given lined black, consistently with 14in. figures. However, scrapping began in 1927 and all had gone by the end of 1932, so it is certain that a fair number were scrapped in their pre-1928 colours. As usual, it is impossible to be certain whether the post-1927 insignia were black or red shaded. We give a 'best-estimate' in our summary.

Livery Samples

Code A1   14656-9/65-9/70§/71§/72§
Code B4   14659§/63/7/70§
Code B7   14664/72

Notes: a) Locomotives Nos. 14659/66-70 have modified cabs with smaller 'cut-away'
b) Locomotives Nos. 14671-2 have Whitelegg boilers/cabs
c) Letter spacing, post-1927, at 53in. centres or slightly more
d) Six wheel tenders marked §

Manson 128 class (LMS Nos. 14673-4; Power Class 3, later 3P)
If James Manson had never designed another engine, these two fine superheated 4-6-0s, built in 1911, would stand tribute to him. Developed out of the 381 class they were carefully modified to accept Schmidt superheaters, and proved outstandingly successful given their overall modest size, which gave but 19,992lb. tractive effort.

They were instantly recognisable from the 381 class by virtue of the larger bogie wheels, longer bogie wheelbase and raised running plate above the piston valves with which they were fitted in succession to the slide valves of the saturated series. In most other respects they followed the 381 class, including the bogie tenders.

We feel sure that if Manson had not retired a year later, the earlier 4-6-0s would have been rebuilt to something like the 128 style, but Peter Drummond took time to be convinced of superheating in spite of the fact that Manson had already conducted trials in 1911 demonstrating a 20 per cent fuel saving with the superheated variant. Fortunately, Whitelegg did not interfere with them, but a class of only two engines stood no chance of survival in LMS days, and they lasted only a year or two beyond the saturated 4-6-0s. They were both painted red in LMS days (Code A1) and No. 14674 is reported to have kept this livery until early 1934. It was then repainted lined black *(Plate 263)* for a few months before withdrawal. No. 14673 went to the scrapyard in late 1933, also wearing lined black livery *(Plate 262)*.

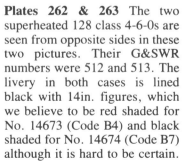

Plates 262 & 263 The two superheated 128 class 4-6-0s are seen from opposite sides in these two pictures. Their G&SWR numbers were 512 and 513. The livery in both cases is lined black with 14in. figures, which we believe to be red shaded for No. 14673 (Code B4) and black shaded for No. 14674 (Code B7) although it is hard to be certain.
A. G. Ellis and Photomatic

Passenger Tank Classes

We have stated elsewhere that the G&SWR did not particularly care for tank engines, and relatively few came to the LMS in the passenger category — fourteen 0-4-4Ts and six 4-6-4Ts. Four of these never received their LMS numbers (Stirling 0-4-4Ts dating from 1879) although allocated LMS Nos. 15241-4. They ran until circa 1925/6, and we give an example in *Plate 264*. The other two classes did receive more attention in LMS days.

Plate 264 Stirling 1 class 0-4-4T No. 728, allocated LMS No. 15241 but never received, seen at Pollockshaws, circa 1923/4.
A. G. Ellis

Manson 326 class 0-4-4T (LMS Nos. 15245-54; Power Class 1, later 1P)
These typically good-looking Manson engines dated from 1893 and were designed for the Glasgow suburban services, working mostly from St. Enoch Shed for the whole of their life. Visually, they were somewhat reminiscent of both the LYR and Midland 0-4-4Ts, and quite unlike the Caledonian breed *(see Chapter 2)*.

Plates 265 & 266 Opposite side views of Manson 326 class 0-4-4Ts Nos. 15246 (ex-G&SWR No. 521) and 15253 (ex-G&SWR No. 528) showing the early red livery, Code A1, and one each of both styles of chimney. No. 15246 was photographed in mid-1931, shortly before scrapping.
A. G. Ellis and Authors' Collection

In 1921, Whitelegg extended the bunkers and gave them coal rails and, in this modified form, all ten reached the LMS — *Plate 265*. By this time, quite a number (but not all) had received replacement cast-iron chimneys of shorter height *(Plate 266)* reducing the overall height from 13ft. to 12ft. 6¾in. We have identified Nos. 15245/8/51/3-4 as having short chimneys in LMS days, and all but No. 15250 of the remainder retaining the taller 'built-up' pattern. Perhaps a reader could help with the missing example. Unlike many classes, the change of bunker and chimney did not spoil the overall lines of the design.

We think the LMS painted them all in red, pre-1928, and we have managed to confirm seven so finished — Code A1. These were Nos. 15245-9/52-3. The class was withdrawn during 1930-32 and we give the only two examples of post-1927 liveries we have been able to locate in *Plates 267 & 268*. We feel certain that a few more were so finished but have no evidence on which to base a valid generalisation.

Plates 267 & 268 These two engines, again bearing each type of chimney, are the only examples we have found of post-1927 liveries on the Manson 0-4-4Ts, although we feel sure that there must have been a few more. Both are in lined black livery and No. 15253 (ex-G&SWR No. 528) undoubtedly carries black shaded transfers, Code B7. No. 15254 (ex-G&SWR No. 529) with 12in. numerals also seems to have plain characters, Code B6, but we are less positive about this one.

A. G. Ellis

Whitelegg 540 class 4-6-4T (LMS Nos. 15400-5; Power Class 5, later 5P)

It was, of course, inevitable that Whitelegg would produce a 4-6-4T, having been partly frustrated in his similar ambition on the LT&SR, but what a 4-6-4T it turned out to be! The concept was quite foreign to the G&SWR and the engines were huge. They were, however, exceedingly well-proportioned, beautifully turned out, and emerged in a welter of publicity in the last independent year of the G&SWR, 1922. Unlike previous G&SWR types, they were usually referred to as the 'Baltics' rather than by their numerical class designation, although according to David L. Smith, 'Big Pugs' was an unofficial title!

Everything about these engines was on a gargantuan scale for the period — not least the price. Again, according to David Smith, the contract price per engine was £16,125 which was money enough, but when one considers that fifteen years later, Crewe was producing Stanier 'Duchesses' at some £10,000 per unit, one cannot help wonder just how Whitelegg got these engines past the G&SWR Board of Directors! Small wonder that the new LMS was not totally impressed.

The 'Baltics' were very handsome engines and, by all accounts, did some good work. Amazingly, for ex-G&SWR men, the crews are said to have liked them, but they had considerable maintenance and servicing problems and were withdrawn in the mid-1930s when, presumably, their original boilers had become life-expired. In this respect, they fared no better than the LYR, FR or LT&SR-designed 4-6-4Ts inherited by the LMS, but perhaps they paved the way for the acceptance on the G&SWR system of the big passenger tank in the shape of the LMS 2-6-4T version, which was to play such an important role in Scotland in later years.

What can be said is that the LMS made some efforts to paint these engines in a distinctive style at all times, consistent with standard practice. They had emerged in full G&SWR livery with polished steel boilers, but the LMS painted them standard crimson (Code A1), in the manner of *Plate 269* and in this guise they probably looked good. One at least *(Plate 270)* received a works grey version of the red livery, (Code A2) but we cannot say whether it ever ran in full crimson with the individual 'LMS'. If so, it was probably the only engine of a Scottish constituent of the LMS which did. We think all were red and have confirmed Nos. 15401-3/5 exactly as *Plate 269*.

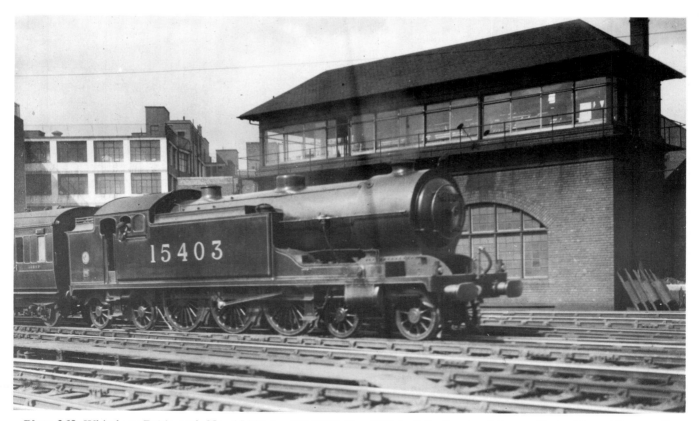

Plate 269 Whitelegg Baltic tank No. 15403 (ex-G&SWR No. 543) in fully lined red livery, Code A1, at Glasgow St. Enoch) circa 1926.

A. G. Ellis

Plate 270 No. 15402 (ex-G&SWR No. 542) in the works grey version of LMS red livery Code A2, a very rare livery in Scotland at this time.

Authors' Collection

After the livery change, the 'Baltics' received lined black, but several of them came out in the so-called 'St. Rollox' style, with large tankside numerals — see *Volume One, page 216*. It suited their imposing lines very well *(Plate 271)*. We have confirmed at least three so finished (Nos. 15401/4-5) and there were possibly more.

Plate 271 Lined black St. Rollox style livery was carried by at least three 4-6-4Ts, of which No,. 15405 (ex-G&SWR No. 545) was one. The livery code is B1 and the insignia were, of course, black shaded with this variant.

Photomatic

Plate 272 This well-known official view clearly shows the orthodox post-1927 lined black livery on 4-6-4T No. 15401 (ex-G&SWR No. 541). The insignia are almost certainly black-shaded, Code B7.

BR (LMR)

Finally, the conventional post-1927 lined black livery was applied *(Plate 272)* and it 'sat' on the engines very well using 14in. Midland numerals. We are fairly certain that black shaded insignia (Code B7) were used and have confirmed all except No. 15403 so finished. We see no reason to believe it was any different.

The engines were scrapped during 1935-7, the last to go being No. 15405, in August 1937.

Chapter 6
Glasgow and South Western Railway — Freight Classes

Freight Tank Classes

The G&SWR was not much more wildly enthusiastic about freight tanks than it was about the passenger examples of the genre, but, over the years it managed to acquire or build rather more than it did on the passenger side. Consequently, by the time of the Grouping, there were 62 in service. Most of these were either 0-6-0Ts or 0-6-2Ts, but there were a few four-coupled examples and we deal with all of them, generally in ascending LMS number order. As befits the nature of the G&SWR, many of its freight tanks were distinctly unique to itself!

0-4-0T/ST — all varieties (LMS Nos. 16040-51)
These twelve engines represented five different types, of which six were representative of one class and the rest either one-offs or matching pairs. We deal with these small groups first.

LMS Nos. 16040-1 (ex-G&SWR Nos. 659, 658) These two 0-4-0ST engines, both illustrated in *Plates 273 & 274*, dated from 1881, having been built by Andrews, Barr and Company for local shunting at Greenock in succession to a pair of older and even smaller engines. Early in LMS days, No. 16041 became the Kilmarnock Works shunter, and No. 16040 went to work the harbour branch at Inverness. No. 16041 was scrapped in 1928 but No. 16040 lasted until 1932, by which time it had received the post-1927 LMS insignia, (Code C13) with 40in. letter spacing on the tank side.

Plates 273 & 274 Opposite side views of the two 0-4-0STs Nos. 16040/1 (ex-G&SWR Nos. 659/8) in LMS livery, Code C4. No. 16040 was pictured during its time at Inverness, the other is at Kilmarnock Works.
A. G. Ellis

LMS No. 16042 (ex-G&SWR No. 734) This engine never received its LMS number and was an Andrew Barclay 0-4-0ST dating from 1883. It was acquired (by purchase) for the G&SWR in 1885 and by 1894 it had become the Kilmarnock Works shunter *(Plate 275)*. It performed this task until scrapped in 1925 and replaced by No. 16041 *(above)*.

Plate 275 Andrew Barclay 0-4-0ST No. 734 in its special G&SWR 'Kilmarnock Works' livery during 1923.
A. G. Ellis

LMS No. 16043 (ex-G&SWR No. 735) This Peckett 0-4-0ST, built in 1904, *(Plate 276)* was acquired by the G&SWR when it took over the Ayr Harbour Commissioners' solitary engine in 1919, along with the harbour itself. The LMS gave it a proper livery (Code C4) and in 1929 sent it to Perth (for the traffic to Gleneagles Hotel). It was scrapped in 1930.

Plate 276 Peckett 0-4-0ST No. 16043 (ex-G&SWR No. 735), ex-Ayr Harbour Commission, is seen in very disreputable LMS livery Code C4.
Authors' Collection

LMS Nos. 16050/1 These two engines were 0-4-0STs, built by Neilson in 1887 for the Glasgow and Paisley Joint Railway as Nos. 1, 2. The G&SWR had almost always serviced them and recorded them in its own totals from 1913. They are believed to have been allotted G&SWR numbers 736 and 737 in 1919 but there is no evidence that they ever received them, neither did they get their LMS numbers, and they were withdrawn in 1924 *(Plate 277)*.

Plate 277 G&P Jt. Rly. No. 1 was one of two Neilson 0-4-0STs eventually owned by the G&SWR. It should have become LMS No. 16050 but never did.
Authors' Collection

Manson 272 class 0-4-0T (LMS Nos. 16044-9) These six engines, built from 1907-9, were quite extraordinary machines — the heaviest 0-4-0Ts (at some 39½ tons) ever to operate in Britain, yet packed on to a 7ft. 6in. wheelbase. They were designed for use at Greenock and Ardrossan where, especially in the former case, radii were sharp and gradients very steep. In spite of these constraints, Manson produced his usual neat solution to the problem *(Plates 278 & 279)* although the end overhangs were considerable.

Plate 278 This less than clear view does, none the less, emphasise the front and rear overhangs of the Manson 272 class 0-4-0Ts. No. 16046 (ex-G&SWR No. 318) is in LMS livery Code C4, and was photographed at Ardrossan in 1925.
A. G. Ellis

Plate 279 Close-up detail of 272 class 0-4-0T No. 16048 (ex-G&SWR No. 320), livery Code C4.
A. G. Ellis

The LMS renumbered them all and we have confirmed at least three in correct pre-1928 livery (No. 16046/8 — Code C4; No. 16047 — Code C5). The others were probably similar. All but one were scrapped in 1930 and the last example (No. 16049, withdrawn in 1931) is the only one we have confirmed in post-1927 livery, Code C13, with somewhat close-spaced letter centres on the tank side.

Manson 266 class 0-4-4T (LMS Nos. 16080-5)

These distinctive engines were yet another almost uniquely G&SWR concept. Effectively they were a revised version of Manson's 14 class 0-6-0Ts *(below)* but with a trailing bogie to help them get into tightly curved sidings. In this respect, they were designed to meet similar conditions to those which gave rise to the LNWR 'Bissel' tanks — *see Volume Two, Chapter 3.*

As usual, Manson produced a very neat design — in essence identical, mechanically, to the 14 class 0-6-0Ts, but the short wheelbase plus outside frame bogie was distinctly unusual. They also had steam sanding and brakes, and had vacuum brakes and train pipes. Eventually they all ended up at Ayr and Ardrossan for shunting duties. *Plates 280 & 281* show both sides of the design.

Plates 280 & 281 Opposite side views of Manson 266 class 0-4-4Ts Nos. 16083 (ex-G&SWR No. 308) and 16085 (ex-G&SWR No. 310) both bearing livery Code C4. Note the similarity with the 0-6-0T type in *Plate 282*.
A. G. Ellis and Authors' Collection

In LMS days they should all have received the freight livery (Code C4) shown in our illustrations, but No. 16080 (withdrawn 1925) was never renumbered. We can confirm Nos. 16081/3-5 as being so finished but we have no records of any post-1927 liveries. Since the final five were withdrawn during 1930-2, one or two may have been repainted.

Manson 14 class 0-6-0T (LMS Nos. 16103-17) These engines represented what David L. Smith has called 'one more timid essay' into the field of tank engines when Manson built the first four in 1896 for yard shunting. Six and five more were added, respectively, in 1903 and 1914. They displayed the usual neat Manson lines and showed more than a passing resemblance to Johnson's similar 0-6-0Ts for the Midland Railway — *see Volume Four*. Indeed we find ourselves wondering, not for the first time when considering Manson's engines, whether there was any 'linkage' between the two railways in locomotive matters. They were operating allies, used similar carriage colours and, mechanically, their carriage running gear in the late 1890s was much the same. They had, in fact, tried to amalgamate and Manson's period of office (in his earlier years) coincided with that of Johnson in his finest hours. Perhaps we are too fanciful!

These engines were probably the nearest the G&SWR came to an orthodox freight tank, and the LMS received all fifteen of them, most as shown in *Plate 282*. However, by that time, Whitelegg had (naturally) rebuilt a few — on this occasion with extended side tanks, and larger bunkers *(Plate 283)*. These were LMS Nos. 16105/9/13/7. These four also received a shorter, but not unshapely, cast-iron chimney.

Plate 282 Manson 14 class 0-6-0T No. 16107 (ex-G&SWR No. 284), livery Code C1, virtually as built.

BR (LMR)

Plate 283 Whitelegg 14 class 0-4-0T rebuild No. 16113 (ex-G&SWR No. 275) in dirty pre-1928 livery, Code C1. Note, apart from the changed physical details, the unfortunate positioning of the figures.

Authors' Collection

Apart from using 18in. figures, pre-1928 (not uncommon in Scotland) where 14in. might have been better, the engines were straightforward, in livery terms, and we give a fair summary *(below)*. However, there appeared to be some confusion about which part constituted the vertical centre of the 'elongated tank' variety which caused some untidiness of letter spacing on some engines *(e.g. Plates 283 & 284)*. It was, eventually, resolved — *Plates 285 & 286*.

Apart from No. 16103, of which we have never seen a picture and which was withdrawn in 1928, the rest of the series was scrapped between 1930 and 1932.

Plate 284 This Whitelegg 14 class rebuild, No. 16109 (ex-G&SWR No. 286) has post-1927 livery with 10in. figures, Code C13, but the 'LMS' placement looks peculiar.

A. G. Ellis

Plate 285 The balance of letter positioning on No. 16117 (ex-G&SWR No. 283) is much better, and the 14in. figures, Code C15, enhance the effect.

BR (LMR)

Plate 286 Unrebuilt 14 class 0-6-0T No. 16115 (ex-G&SWR No. 279) shows the most common post-1927 livery style, Code C14, for these engines.

Authors' Collection

Livery Samples

Code C1   16105§/6-8/10/13§/4/6
Code C13   16109§
Code C14   16104/5§/11/15
Code C15   16117§

Notes: a) Whitelegg rebuilds marked §
        b) Letter spacing generally about 40in. centres — post-1927

Drummond 5 class 0-6-0T (LMS Nos. 16377-9; Power Class 2, later 2F)
These three chunky tank engines were introduced by Drummond in 1917 for heavy duty at Greenock and Ardrossan, in succession to elderly 0-4-0 tender engines of Patrick Stirling vintage. The 0-6-0Ts had flangeless centre wheels to help them round the sharp curves, and proved very satisfactory.

The LMS received all three and, we believe, painted them as shown in *Plate 287*. However, at a later date, No. 16378 received an extraordinary and, as far as we know, unique combination of bunker panel and 12in. standard post-1928 numerals — *Plate 288*. No. 16377 was scrapped in 1932 and the other two were sold out of service in 1934. Happily, one of them (No. 16379) when finally withdrawn in 1962 from Llay Main Colliery (North Wales) was 'spotted' and is now preserved at the Glasgow Transport Museum as the only ex-G&SWR engine to survive. While it is by no means the most typical of G&SWR engines, it is not entirely unfortunate that the G&SWR should be represented by one of the later Drummond designs. In spite of all that we, and others, have said about this much misunderstood man, his final G&SWR efforts were by no means inferior machines and, in the event, well outlasted the designs of his predecessors.

Plate 287 Drummond 0-6-0T No. 16379 (ex-G&SWR No. 324) at Ardrossan, in April 1928, in orthodox early LMS livery, Code C4, but employing small-sized bunker panel. This engine is now preserved in Glasgow Transport Museum.

A. G. Ellis

Plate 288 We think this is unique. It shows 0-6-0T No. 16378 (ex-G&SWR No. 323) in pre-1928 style livery but employing 12in. high post-1927 type transfer figures. We have no code system for this one and have not encountered another single LMS engine painted this way. The bunker 'panel' seems to be the normal size version, fighting for space with the end sections missing. This is, perhaps, a hand-painted 'special'.

Authors' Collection

Drummond/Whitelegg 45 and 1 class 0-6-2Ts (LMS Nos. 16400-27, later 16900-28; Power Class 3, later 3F)

By 1915, when these engines were introduced, Peter Drummond seems to have worked himself into a somewhat more acceptable position than when he took over from Manson. Big tank engines were not quite what the G&SWR was accustomed to getting, but on the steeply-graded branches in the Ayrshire Coalfield, they carried out some excellent work and the men did become versed in their handling.

Eighteen of the 45 class were built during the Drummond period (between 1915 and 1917) and ten more were on order when Whitelegg took over. All he did was to increase the tank capacity, modify the cab 'cut-out' and other minor details, and put the driver back at the right-hand side! These were the 1 class variants of 1919 but, in either form, the engines were neat and well-proportioned *(Plates 289 & 290)*. For some reason, the LMS numbered the Whitelegg engines ahead of the Drummonds (Nos. 16400-9 and 16410-27 respectively) and when they were renumbered 16900-27 to make room for LMS Class 3F 0-6-0Ts, this 'wrong way round' sequence was maintained.

Plate 289 No. 16423 (ex-G&SWR No. 24) shows one of the first Drummond series of 0-6-2Ts in early LMS livery, Code C1, with its first series LMS number. It later became No. 16923.
W. T. Stubbs Collection

Plate 290 No. 16902 (ex-G&SWR No. 3) shows one of the Whitelegg period 0-6-2Ts in the short-lived unlined version of the St. Rollox livery style, employed to use up spare 18in. transfers, circa 1930. Note the higher tanks, changed cabside cut-out and somewhat more rounded dome — a touch of the Tilbury line perhaps?
Photomatic

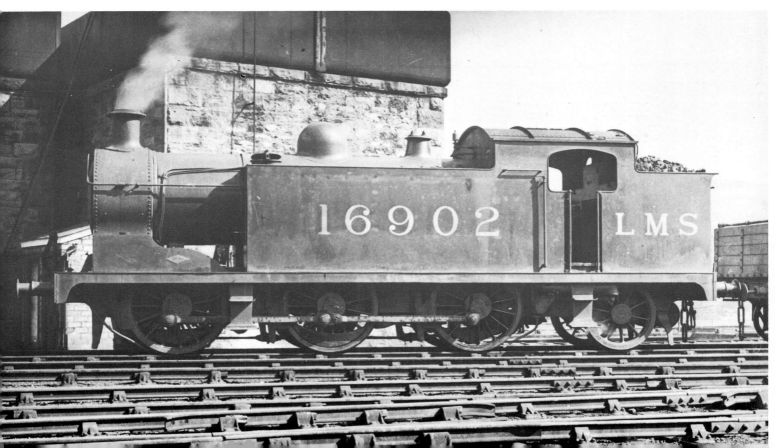

Plate 291 No. 16910 (ex-G&SWR No. 11) the first of the entire class, shows the alternative pre-1928 livery, Code C2, with rounded panel.

Authors' Collection

The general design was based on the 0-6-4T designed by Drummond for the Highland Railway *(Chapter 7)* and there were many points in common. The LMS found plenty of work for them to do and moved them to widely scattered parts of the system over the years. None were scrapped until 1936, and one of them (LMS No. 16905) held the distinction of being the last ex-G&SWR engine in capital stock. They went a long way to redeem Drummond's reputation.

Appearances hardly varied and our pictures and livery summary give a good cross-section of styles. Post-1927, livery was very consistent indeed *(Plate 292)* and we have only found one wartime repaint *(Plate 293)*.

Plate 292 Post-1927 style, with 14in. figures, Code C15, was the most common livery style for the 0-6-2Ts. No. 16906 (ex-G&SWR No. 7), a Whitelegg-built example, is featured here.

A. G. Ellis

Plate 293 The last ex-G&SWR engine in capital stock, 0-6-2T No. 16905 (ex-G&SWR No. 6), just reached BR in the form shown here, wearing wartime insignia, Code C21.
Authors' Collection

Livery Samples

Code C1   16400/23-4; 16905/14/8
Code C2   16910-2
Code C3   16902/16/18/24 (all with 14in. LMS on bunker side — *as Plate 290*)
Code C15   16900-1/3-4/6-9/12-5/7/9-22/5-7
Code C21   16905

Note: 'LMS' letter spacing — post-1927 — generally 40in.-53in. between centres

Freight Tender Classes

Like every other collection of locomotives owned by this distinctive company, the freight tender engines were characterised by their infinite variety. At the time of the Grouping they ranged from some almost primeval relics from Patrick Stirling's period, to the heaviest 0-6-0s to be seen in Britain up to that time. There was also a class of heavy 2-6-0 engines — a rare wheel arrangement in Britain for the period — while another distinctive feature was the considerable number of 0-4-2 tender engines in service, 55 in all, again an almost uniquely G&SWR situation as far as the constituents of the LMS were concerned.

Stirling 0-4-2 types — all classes (LMS Nos. 17021-75; Power Class 1, later 1F)
The 0-4-2 tender type was something of an old fashioned approach to freight working, even when many of the G&SWR examples were being built; but the Company liked them, developed them, and rebuilt them, in consequence of which, a considerable number survived until 1923. These were in four classes, introduced variously between 1866 and 1874.

a) P. Stirling 141 class (LMS Nos. 17021-2) Two late survivors of Patrick Stirling's final 0-4-2 design of 1866, retained to work the Mayfield mineral branch, just reached the LMS, but were scrapped in 1923 and 1925 respectively, and we have no illustration to offer of either of them. Neither was renumbered.

b) J. Stirling 208 class (LMS Nos. 17023-7) These five engines, all withdrawn in 1923, dated from 1873. As built they were very similar to the earlier 187 class *(below)*, but because they had not been extensively modified by Manson, they were withdrawn earlier, without receiving LMS numbers. Again we can offer no view of this type.

c) J. Stirling 187 class (LMS Nos. 17028-34) James Stirling's 187 class was introduced in 1870 as a direct continuation of his brother's work, but with larger wheels. None survived into LMS days 'as built', but during 1900/1 Manson had rebuilt seven with new cabs and his own domeless boilers. Classed as complete renewals, these seven did survive to the Grouping and three were renumbered. Two of the seven had detachable tender cabs for the Whithorn branch *(Plates 294 & 295)*. By LMS times, some, at least, were fitted with more modern tenders, both with and without weatherboards *(Plates 296 & 297)*. These engines illustrated are the only three examples which were ever given LMS numbers and liveries and were the last to go, in 1930.

Plates 294 & 295 Opposite side views, with and without tender cab, of Stirling 187 class 0-4-2 No. 17029 (ex-G&SWR No. 269), livery Code C1, still coupled to a Stirling tender.

A. G. Ellis and Photomatic

Plates 296 & 297 These two 187 class 0-4-2s, No. 17031 (ex-G&SWR No. 271) and No. 17032 (ex-G&SWR No. 272) are coupled to replacement tenders, that of No. 17031 having a weatherboard. No. 17032 has the alternative cab panel, livery Code C2, and the other engine is livery Code C1.
Authors' Collection

d) J. Stirling 221 class (LMS Nos. 17035-75) The 221 class, originally sixty strong, was the second most numerous type on the G&SWR, and was James Stirling's final 0-4-2 design, dating from 1874. They were somewhat larger than the 187 and 208 classes and became the G&SWR 'standard' goods engines for some thirty years. They were considered good enough for Manson to completely rebuild thirty from 1901-3, and fit them with new cabs and domed boilers. These thirty all came to the LMS but, of the original type *(Plate 298)* only eleven reached the LMS (Nos. 17035-45). All of this group were withdrawn without renumbering.

Plate 298 This picture of 221 class 0-4-2 shows the original Stirling design and depicts G&SWR No. 637 (allocated LMS No. 17037) at Hurlford in 1927, by which time it was the last unrebuilt survivor.
A. G. Ellis

The Manson rebuilds (Nos. 17046-75 — *Plate 299*) fared much better, a modest majority being renumbered, and many lasted until the late 1920s. As with other Stirling designs, some of them received replacement tenders from scrapped engines of Smellie's 2-4-0 series *(Plate 300)*. We have confirmed at least a dozen in correct pre-1928 livery *(below)* and one, at least, lasted long enough to get the post-1927 treatment. It was one of the last two survivors, both withdrawn in 1931 *(Plates 301 & 302)*. Those confirmed as not renumbered were Nos. 17049-52/8/62-3/8-70/3-5.

Plate 299 (Above) Stirling 221 class 0-4-2 No. 17047 (ex-G&SWR No. 243) as rebuilt by Manson, LMS livery Code C1.

Authors' Collection

Plate 300 (Below) No. 17066 (ex-G&SWR No. 263), fitted to a replacement tender designed by Smellie. There is no evidence of LMS ownership except the numberplate.

R. S. Carpenter Collection

Plates 301 & 302 The last surviving 221 class engines were Nos. 17066 and 17067 (ex-G&SWR Nos. 263/265). No. 17066 is in post-1927 livery, Code C13. No. 17067 (with cab weatherboard) is wearing livery Code C2.
A. G. Ellis and Authors' Collection

Livery Samples

Code C1	17047/8§/53§/4§/7/9§/60-1/4-5/72§
Code C2	17067
Code C4	17048§
Code C13	17066§

Note: Replacement tenders marked §

Stirling 0-6-0 types — all classes (LMS Nos. 17100; 17103-11)

A handful of elderly 0-6-0s from the Stirling era just managed to reach the LMS. The oldest, allocated LMS No. 17100, was the last survivor of Patrick Stirling's 58 class of 1866 *(Plate 303)*. David L. Smith records that for some reason it was repainted in 1919 and sent to Dumfries where it was greeted with such surprise that they could find no work for it to do! It was scrapped in 1923 without being renumbered.

Plate 303 Patrick Stirling 58 class 0-6-0 No. 164A, later GSWR No. 619 in G&SWR colours, was the last survivor of its kind, and should have become LMS No. 17100, but did not do so.

Authors' Collection

The other nine engines (Nos. 17103-11) were from James Stirling's 13 class *(Plate 304)* which dated from 1877. It was his first and only G&SWR 0-6-0 type, developed from his 221 class 0-4-2s but, unlike the latter, Manson did not rebuild the 0-6-0s. The design is possibly of most interest in having been adopted by the South Eastern Railway, whence James Stirling moved in 1878. None of the 13 class are believed to have been renumbered by the LMS, except perhaps the last survivor (No. 17109, ex-G&SWR No. 577), which was scrapped in 1928. They were James Stirling's last tender engines for the G&SWR (his 0-4-4Ts did not appear until after he had left) and, like all his engines, were universally popular.

Plate 304 James Stirling 13 class 0-6-0 No. 578. This locomotive was allocated LMS No. 17110 but it was never carried.
Authors' Collection

Smellie 22 class 0-6-0 (LMS No. 17112-64/17203-12; Power Class 2, later 2F)

This design, Hugh Smellie's only 0-6-0 type, was the most numerous class of engine on the G&SWR, and was built continuously from 1881 to 1890. The engines were similar in general layout to his 2-4-0s, and quite a number came down to the LMS, largely unmodified *(Plate 305)*. As with other designs by Smellie, Manson fitted quite a number (twelve in all) with small domed boilers of similar size *(Plate 306)*. These two types of boiler tended to get changed between engines at shopping times, but we note the variant in our summary *(below)*.

Towards the end of the Manson period, a further ten were given larger and higher-pitched domed boilers but still retaining their rounded cabs *(Plate 307)*. This was classed as a complete renewal and therefore the LMS regarded them as newer than the Manson 0-6-0s proper *(below)* and numbered them separately as Nos. 17203-12.

Plate 305 The harmonious lines of Smellie's 22 class 0-6-0 design are well-shown by No. 17142 (ex-G&SWR No. 612) in pre-1928 LMS colours, Code C1. Note the lack of front coupling links — a common enough feature on many G&SWR freight tender engines but very obvious in this view.

BR (LMR)

Plate 306 Manson's small-domed boiler rebuild of a 22 class 0-6-0 — No. 17117 (ex-GSWR No. 139) in early LMS livery, Code C2.

A. G. Ellis

Plate 307 The later Manson rebuilt form of the 22 class is shown by No. 17203 (ex-G&SWR No. 140) — livery Code C1.

Authors' Collection

Plate 308 The Whitelegg rebuilding transformed the appearance of the 22 class — well-illustrated here by No. 17138 (ex-G&SWR No. 608), livery Code C1. They looked attractive, but . . .!

BR (LMR)

Not surprisingly, and to their detriment, Whitelegg's modifications with X2 boilers *(see Chapter 5)* were also applied, but only to ten examples, and not until 1924-5 *(Plate 308)*. These were LMS Nos. 17118/36/8/41/4/6/54-6/63. Thus, four variations were to be seen on the class by the mid-1920s (small domeless boiler, small-domed boiler, larger-domed boiler, Whitelegg style). Regardless of type, these engines did not fare any better than most ex-G&SWR classes. Most had gone by 1932 but one or two lingered on, the last survivor — a Whitelegg rebuild No. 17163 — lasting until 1935.

A considerable number did not receive LMS numbers and very few received the post-1927 livery. Before the livery change, however, many received the correct form, universally with 18in. figures. Confirmed as not being renumbered are No. 17112-6/21/3/6/8/30/2/4-5/7/43/5/8-52/7/9-62/4.

Livery Samples

Small boilered type
Code C1   17119/25/31§/3/42/58
Code C2   17117§/29/53§
Code C14   17122§ (53in. letter spacing)

                    Domed version marked §

Larger Manson boiler, rounded cab
Code C1   17203/8/10
Code C2   17203

Whitelegg rebuilds
Code C1   17118×/38/41/56
Code C2   17136/44
Code C13   17154 (40in. letter spacing)
Code C15   17138/63

            × *No cabside marking*

Smellie/Manson 306 class 0-6-0 (LMS Nos. 17165-84; Power Class 2, later 2F)
These twenty 0-6-0s were on order in 1892 when Manson took over, and he had slight alterations made before they entered service, including higher pressure, domed boilers with rather more and slightly smaller tubes than the original version by Smellie. The round-cornered cab was retained, however, and all twenty reached the LMS essentially unmodified — *Plate 309*.

After Grouping during 1924/5, three of them (LMS Nos. 17167/70/6) were given the Whitelegg cab and X2 boiler — *Plate 310*. In this form they were identical to the 22 class rebuilds *(above)*. Four are known not to have been renumbered (Nos. 17165/78/80/4) but those which were, seem to have received conventional liveries. A few lasted until 1932 and some, as illustrated here, achieved the post-1927 livery.

Plate 309 The 306 class 0-6-0 was virtually Manson's modification of Smellie's 22 class. No. 17174 (ex-G&SWR No. 187) is shown in post-1927 livery, Code C14, still remaining very much in 'as built' condition.

                                                            BR (LMR)

Plate 310 No. 17176 (ex-G&SWR No. 189), in LMS livery Code C14, was one of only three Whitelegg style rebuilds of the 306 class — the visual transformation was just as thorough as with the earlier designs.

A. G. Ellis

Livery Samples

Code C1 17166/9/70§/1-2/4/7/83×
Code C14 17173-4/6§ (all with about 53in. letter spacing)

Note: Whitelegg style rebuilds marked §
  × *No cabside marking*

Manson 160 class 0-6-0 (LMS Nos. 17185-202; Power Class 2, later 2F)
This next group of eighteen 0-6-0s were, essentially, the 'pure' Manson version of the 306 class with his own design of cabs. The boilers sat very low relative to the cab roof, and some did not like their proportions, but we cannot really see why — *Plate 311*. Chimneys were replaced on at least one of them *(Plate 312)* and probably on more.

Three examples are confirmed as never receiving LMS numbers (Nos. 17186/96/200). One of these (No. 17196) was sold in 1926 for colliery service in Northumberland. It survived there until 1953 when it was broken up at Seaton

Plate 311 The Manson 160 class was considered by some to be a little 'hump-backed' and low-boilered in appearance. We cannot quite see why. No. 17188 (ex-G&SWR No. 163) is in livery Code C1.

BR (LMR)

Plate 312 This Manson 160 class 0-6-0 should have become No. 17196, but never did. It is pictured as G&SWR No. 171 with replacement chimney.
Authors' Collection

Delaval — a few years later and it may well have been saved. In the same year that No. 17196 was sold, four more were given the Whitelegg X2 boiler treatment (LMS Nos. 17190/5/201-2) — *Plate 313*.

Their LMS liveries were orthodox but, as usual, few survived to receive the later style and the last withdrawal occurred in 1933.

Livery Samples

Code C1   17185/7-9/92/5§/8-9/201/2§
Code C14   17195§ (with 40in. letter spacing)

Note: Whitelegg style rebuilds marked §

Plate 313 Like most Whitelegg style rebuilds of G&SWR 0-6-0s, the 160 class was actually dealt with in early LMS days, so No. 17201 was never a G&SWR engine in the form shown here (LMS livery Code C1). It came to the LMS as G&SWR No. 176 in the form seen in *Plate 311*.

Stephen Collection, courtesy NRM

Plate 314 Locomotives of the 361 class were possibly the best looking Manson 0-6-0s, well-shown here by No. 17481 (ex-G&SWR No. 127), livery Code C1, one of the rather small proportion which remained unrebuilt after 1925. The apparent front coupling *(see caption, Plate 305)* is actually a set of 'loose' 3 links!

Stephen Collection, courtesy NRM

Manson 361 class 0-6-0 (LMS Nos. 17474-507; Power Class 2/3, later 2F/3F)

The 361 class engines began to emerge soon after the end of construction of the 160 class. When compared with the 160 class, the wheelbase was extended, the boiler height raised, and the new design was a typically well-proportioned Manson creation. All told, 34 were built between 1900 and 1910 and, until 1920, all remained substantially the same and were highly regarded and successful engines *(Plate 314)*.

In 1920, a start began, inevitably, to the fitting of Whitelegg X2 boilers and cabs on many of them. This rebuilding went on until early LMS days (1925) by which time, 21 of the 34 had been converted *(Plate 315)*. The rebuilds were classed 3/3F and the originals 2/2F by the LMS. As usual, Whitelegg produced a handsome enough style, but the rebuilds were never as good as the originals. Those engines remaining unrebuilt after 1925 were Nos. 17474-8/80-1/4/92-6/506-7. The class lasted well into the 1930s, the last to go being in 1937 — a Whitelegg rebuild (No. 17497). One of the unrebuilt engines (No. 17494) although not withdrawn until 1928, failed to be renumbered.

Visually, the two types changed but little during LMS days, but there was one point of difference on the last two examples of the unrebuilt series (Nos. 17506-7). These were built later (1910) and given the final pattern Manson six wheel tenders *(Plate 316)* which they retained, along with the taller pattern cab of the later 17 class *(below)*.

LMS liveries were straightforward and, for a change, quite a number received post-1927 styles. The 12in. figures were favoured but we have confirmed examples of all three sizes *(Plates 317 to 319)*.

Plate 315 The Whitelegg style rebuilt version of the 361 class, No. 17479 (ex-G&SWR No. 124) is seen here in the alternative form of pre-1928 livery, Code C2.

Authors' Collection

Plate 316 The last two class 0-6-0s received the later and taller Manson cab and the new style tender which together did not look quite as good. No. 17506 (ex-G&SWR No. 101) is seen in livery Code C1, leaving Glasgow (St. Enoch) with a stopping train whose first vehicle is an ex-MR lavatory clerestory coach. It is a lovely evocative picture of a lost age!

A. G. Ellis

Plates 317 to 319 Three sizes of post-1927 insignia are shown on 361 class 0-6-0s in these views — Codes C13, C14 and C15 respectively. The engines are unrebuilt No. 17476 (ex-G&SWR No. 118) and rebuilt Nos. 17485 (ex-G&SWR No. 134) and 17504 (ex-G&SWR No. 113).
Authors' Collection, W. L. Good and A. G. Ellis

Livery Samples

Original Manson design (see above regarding Nos. 17506-7)
Code C1   17474/7/81/96/506
Code C2   17475/8
Code C13   17476 (53in. letter spacing)
Code C14   17507 (53in. letter spacing)

Whitelegg rebuilds
Code C1   17482-3/6-7/9-90/500/4
Code C2   17479/83/503
Code C14   17485/90/2-3/501 (40in.-53in. letter spacing)
Code C15   17483/97/504 (53in. letter spacing)

Manson 17 class 0-6-0 (LMS Nos. 17508-22 plus 17523-4; Power Class 3, later 3F)
The 17 class was Manson's final 0-6-0 design, dating from 1910 and, as with his 4-4-0s *(Chapter 5)* he tried the idea of using a larger boiler with an existing design of chassis. Thus, the 17 class was basically a large boilered 361 class, using the 240 class 4-4-0 boiler. They were all classed as 3/3F by the LMS and *Plates 320 & 321* show both sides of the original version, coupled to the flush-sided tender. All were renumbered in or after 1923.

Plates 320 & 321 The 17 class Manson 0-6-0s, with larger boilers, were not regarded by some as being visually so harmonious as his earlier types. They certainly had smaller domes and attenuated chimneys, but they were still very neat engines. Nos. 17513 (ex-G&SWR No. 91) and 17516 (ex-G&SWR No. 94) are shown here, both wearing pre-1928 livery Code C1.

BR (LMR)

Whitelegg rebuilt only one of them (No. 17509) and, unusually, retained the original tender but added coal rails *(Plate 322)*. Two new engines to the Whitelegg variant were built in 1921 but received Whitelegg's more usual tender. These were LMS Nos. 17523-4 *(Plate 323)*. In due course, some slight changes took place to the main series, perhaps the most obvious being a change in the smokebox door *(Plates 324 & 325)*. We do not know how many were thus treated — probably most of the longer term survivors. The engines lasted generally to the early/mid-1930s, and the last withdrawal was in January 1937 (No. 17521).

Plate 322 The unique Whitelegg style 17 class rebuild retaining original tender, No. 17509 (ex-G&SWR No. 87), livery Code C1.

A. G. Ellis

Plate 323 No. 17523 (ex-G&SWR No. 150) was built new to Whitelegg's rebuilt form of the 17 class — LMS livery Code C1. Note the Whitelegg style tender.

A. G. Ellis

Plates 324 & 325 Two different styles of smokebox door replacement are visible on 17 class 0-6-0s Nos. 17512 (ex-G&SWR No. 90), livery Code C14, and 17519 (ex-G&SWR No. 97), livery Code C15. We would like to know more of this feature if readers can help.
*Photomatic
and M. J. Robertson Collection*

The LMS painted them in orthodox fashion and our summary contains no unusual variants. We suspect that they all received the early livery, but since both types of cab panel were used, we cannot say how many of each might have been seen. After 1927, 12in. figures were the favoured, but not exclusive, solution.

Livery Samples

Code C1	17508/9§/13/6/8/20/3§/4§
Code C2	17514-5
Code C14	17512/5/7/21 (40in.-53in. letter spacing)
Code C15	17519/24§

Drummond 279 class 0-6-0 (LMS Nos. 17750-64; Power Class 4, later 4F)

These were the engines which caused all the trouble — Drummond's first design, introduced in 1913, wherein every cherished G&SWR precept was thrown out of the window! They were Britain's heaviest 0-6-0s at that time, were woefully slow, and full of features which the men either did not understand or did not want to. They would steam, as long as the firemen could bale in coal as if it was going out of fashion — one hundredweight per mile was not unheard of! The driver was put on the 'wrong' side of the cab and no one really liked them.

No doubt their faults were exaggerated but, by any standards, they were not particularly good. In later years, certain features were removed, (including their feed-water pumps which had earned them the nickname 'Pumpers') and they did some reasonable work. For all that, they lasted no longer than many older engines and were all withdrawn between 1930 and 1933, unloved to the end!

Yet, for all their faults, they had a handsomeness of line which was shared by most of Peter Drummond's engines for the G&SWR, given the overall massiveness of the design. Although he was inspired, self-evidently, by brother Dugald's LSWR efforts, Peter Drummond seems to have had some good draughtsmen working for him to 'clean up' the often ponderous and clumsy lineaments which many contemporary LSWR engines displayed.

Plate 326 Drummond 0-6-0 No. 17756 (ex-G&SWR No. 77), very clean in pre-1928 livery, Code C1, shows off its massively well-proportioned lines in this view, to good effect.

A. G. Ellis

We give an early LMS picture of the type in *Plate 326*. They never changed significantly and such few liveries as we have located are listed below.

Livery Samples

Code C1	17756/9	
Code C2	17760/2	
Code C13	17753	} letter centres at about 53in.
Code C14	17750/8/64	

Drummond 16 class 2-6-0 (LMS Nos. 17820-30; Power Class 4, later 4F)

This time, Drummond got it almost 100 per cent right with one of the best freight engines ever owned by the G&SWR, albeit beset with niggling maintenance problems and irritating minor defects. By 1915, the virtues of superheating had been realised on Drummond's 4-4-0s *(see Chapter 5)*, so the next batch of main line goods engines was a superheated development of the ill-fated 279 class 0-6-0s. Robinson superheaters were fitted and the extra weight at the front end caused the adoption of a leading pony truck. In this there was a parallel with Caledonian experience *(see Chapter 4)*, but the G&SWR engines at least looked as though they were designed as 2-6-0s, rather than converted as an afterthought. Eleven were built and all reached the LMS.

Known from the outset as the 'Austrian Goods' — for no good reason apart from a totally unsubstantiated belief that some of the material used in their manufacture had been destined for a war-cancelled contract to Austria — the engines were very economical, free running and good on the hills. Moreover, they could accept the appropriate LMS Northern Division standard (i.e. Caledonian) boiler, so they remained in service for longer than most. They were based in Carlisle and, after the Grouping, the LMS used them to Edinburgh, Perth, Dundee and even over the Settle and Carlisle line. However, overheated big ends and other problems caused their reliability to be suspect and they were rarely risked on passenger trains. As boilers finally became worn out, they were gradually withdrawn between 1935 and 1947 — but there was no great carnage.

Like the 0-6-0s, they were massively well-proportioned and their appearance hardly varied. We show opposite side views in *Plates 327 & 328*. LMS liveries were orthodox except for one known 'hiccup' — a very rare thing to happen in Scotland — which we give in *Plate 329*.

Plates 327 & 328 Opposite side views of Drummond 16 class 2-6-0s Nos. 17829 (ex-G&SWR No. 60) and 17826 (ex-G&SWR No. 57), both carrying post-1927 LMS livery, Code C14. No. 17829 is thought to be carrying a CR type boiler — note the different safety-valves.

Authors' Collection and L. Hanson

Plate 329 At the livery change-over, No. 17827 (ex-G&SWR No. 58) had its number stencilled below the cab panel. For the record, this was one of two 16 class 2-6-0s to have worked over the Settle and Carlisle line during 1927/8; the other was No. 17821.

A. G. Ellis

Livery Samples

Code C1   17822/5/6-7
Code C2   17820/9/30
Code C14   17820-30 (the whole class!) } Letter spacing at about 53in. centres
Code C15   17821

Chapter 7
Highland Railway — All Classes

We turn our attention, finally in this volume, to one of the most interesting railways anywhere in the world — the Highland. It is a curious fact that no matter what the individual preferences of our railway enthusiast friends may be — GWR, LNER, Southern or whatever — we have yet to meet anyone who did not have some degree of affection for this far-flung outpost of the LMS. It is hard to see why this should be, and to a large extent it has no relevance to the locomotive story, except perhaps to state that, because of the extreme interest shown in this small railway by photographers and others, we have managed to discover, proportionally, more about certain aspects of its engines than any other pre-group system. Whatever the reasons, it is an astonishing fact that we feel able to state, uniquely in this five-volume survey, that we know how almost every ex-HR locomotive was painted for at least some part of its life during LMS ownership. Our master list of HR engines contains but one totally blank entry — 0-4-4T No. LMS 15017. We cannot even say this for the LMS standard classes!

Whatever else one may say about the Highland it was different. Firstly, it operated main line services on an essentially single track system, and secondly, it took over forty years for its network, as inherited by the LMS, to be totally established.

The oldest part of the Highland Railway, the Inverness and Nairn section, was incorporated in 1854 and opened in 1855. During the next ten years, steady developments in the north of Scotland led to a situation where, by amalgamation of the Inverness and Aberdeen Junction (in succession to the Inverness and Nairn) and the Inverness and Perth Junction (via Forres and Dava Moor) the Highland Railway Company was created in 1865. By 1870, the Dingwall and Skye Company opened its line, worked by the Highland, as far as Strome Ferry, while Wick and Thurso were reached in 1874. The cut off on the Inverness to Perth route (via Carr Bridge) was not completed until 1898, a year after the Highland, having legally taken over the Dingwall and Skye line in 1880, reached the Kyle of Lochalsh, thus establishing the network which survives (in large part) to the present day.

To work this system, the Highland Railway developed a highly distinctive and characteristic fleet of engines, 173 strong, at the Grouping. In achieving this, the Company was fortunate for such a relatively small concern, in having obtained some of the finest locomotive engineers in the land as explained in *Volume One*. Of the 173 engines, 23 were withdrawn before LMS numbers were allocated. These were mostly elderly 4-4-0s of the 'Crewe' type, much favoured by the Highland Railway.

The most significant engineer of the Highland Railway was, probably, David Jones (1870-96) in succession to William Stroudley. Stroudley had built but one new engine during his term of office, but Jones virtually revolutionised matters during the late Victorian period, and his influence lived on into the LMS era. He was succeeded by Peter Drummond who, during his reign on the Highland (1896-1911) generally carried on the good work established by Jones. At this time, Peter Drummond was following the precepts which had been laid down by brother Dugald on the Caledonian *(Chapters 1 to 4)* and in his early days on the LSWR, which is just as well for the Highland since, as has been seen *(Chapters 5 & 6)* when Peter Drummond went from Inverness to the G&SWR at Kilmarnock in 1911 and copied some of Dugald's later LSWR ideas, he tended to become something of a walking 'disaster' area! Drummond's successors on the Highland, Smith and Cumming, had less time to develop their ideas, but their contribution will be considered in due course.

The Jones-Drummond evolution gave to many Highland engines a considerable degree of conformity with the Caledonian practice of the Drummond-Lambie-McIntosh era. A consequence of this was that, in Scotland, the LMS found it possible to use Caledonian type boilers on many ex-HR engines, thus prolonging their useful life. This added a little complexity to the story which we try to unravel in the detailed class sections. On the other hand, it imparted a longevity to Highland designs which was not to be seen on any other part of the LMS which had added such a small total of engines to the fleet. Many ex-Highland designs outlived their far more numerous contemporaries on other parts of the system. A parallel may be drawn with the fitting of the LYR type boilers on the ex-Furness engines *(Volume Two, Chapter 11)* but this was a 'drop in the bucket' compared with the procedures practised north of the border.

Passenger Classes

The Highland's passenger engines were so predominantly of the tender type that we have not felt it necessary to split the tender/tank engines. We merely give a chronological review in LMS number order. However, in passing, it is worth pointing out that, with the exception of the LT&SR *(Volume Four)*, the Highland possessed a higher proportion of engines classified 'passenger' than any other railway eventually absorbed by the LMS. In consequence, the 'Highland Section of the Northern Division', as it became known after 1923, became a veritable riot of red engines during the 1923-28 period, as we shall explain.

Jones 'Crewe type' 4-4-0s (LMS Nos. 14271-85; Power Class 1, later 1P)
In the latter part of the nineteenth century, the Highland Railway had become possibly the most widespread user of a type of locomotive which could trace its origins back as far as the crucial developments at Crewe in the 1840s. Frequently,

and erroneously referred to as 'Allan-framed', the basic concept owes far more to the pioneering work of W. B. Buddicom and Francis Trevithick than it does to Alexander Allan, who was merely Works Manager at the time and who, later, claimed for himself far more credit than was strictly due to him. The basic philosophy is encapsulated for all time in the preserved *Columbine* at the National Railway Museum, and proved particularly popular in Scotland, nowhere more than on the Highland Railway.

The fundamental concept was that of a massive double frame at the front holding the outside cylinders, slide bars and cross-head, combined with conventional inside frames to the driving and trailing wheels. This idea had been applied to 2-2-2, 2-4-0 and 4-4-0 types (both tender and tank) and gave a very distinctive visual character to the 'front end' of the engine. As developed and enlarged from 1874 onwards by Jones, it was virtually a trade mark of the Highland system for almost half a century. By LMS days, the type was, not surprisingly, finding itself overtaken by more modern developments, but there were still fifteen survivors which had enough life in them to be given LMS numbers. These were the remaining 6ft. 3in. driving wheel engines of the 'Bruce' and 'Strath' classes of 1886/1891 respectively, and the residual (5ft. 3in.) small-wheeled Skye bogies dating essentially from 1893, but with one example whose origins went back to 1882. In LMS number order the were:

- a) Nos. 14271-6: 'Strath' class, 6ft. 3in. engines built 1892
- b) No. 14277: Original Class L 5ft. 3in. 'Skye Bogie' of 1882
- c) No. 14278: 'Bruce' class, 6ft. 3in., built 1886 — forerunner of 'Strath' type with smaller boiler
- d) Nos. 14279-85: Class L, 5ft. 3in. 'Skye Bogies' built 1893-1901 (the last four by Drummond)

No. 14277 was numbered out of sequence in 1923 in the mistaken belief that it was another 'Bruce' class survivor. Three of them (Nos. 14273/80-1) never received LMS numbers, being withdrawn between 1925 and 1926. The two driving wheel sizes were readily distinguishable *(Plates 330 & 331)* but it is worth mentioning that the distinctive tall louvred Jones chimney was not present on No. 14278 or, originally, on Nos. 14282-5. The former had received a replacement Drummond chimney *(Plate 332)* and the last 'Skye Bogies' had this type from new. However, No. 14283 sported a tall Jones chimney in early LMS days *(Plate 333)*.

Plate 330 A three-quarter rear view of No. 14272 *Strathdearn* (ex-HR No. 92A) at Forres in 1928, showing tender detail and lining layout with red livery, Code A1.
H. C. Casserley

Plate 331 'Skye Bogie' No. 14284 (ex-HR No. 34) in full red livery — Code A1. Note how the driving wheel sandbox projected well above the splasher top on these 5ft. 3in. engines compared with the 6ft. 3in. type.
BR (LMR)

Plate 332 No. 14278 (ex-HR No. 82A), formerly called *Fife* was the last surviving 'Bruce' class 4-4-0, and shows the replacement chimney — LMS livery Code A1. Careful comparison with *Plate 334* will reveal the slightly larger boiler of the latter, compared with the 'Bruce' class.

Photomatic

Plate 333 Although this is an earlier view of No. 14283 than that of No. 14284 *(Plate 331)* it shows the tall Jones chimney which replaced the original Drummond type on this particular engine.

Gavin Wilson Collection

Within but a short time after the Grouping, most of the engines were resplendent in the fully-lined crimson livery, Code A1, and all are believed eventually to have carried it. At this time the Highland section was noticeable throughout Britain for the cleanliness of many of its engines and most of this group were commendably well turned out as many pictures indicate. We have failed to confirm only one of the twelve renumbered engines (No. 14274) in this livery, but since it was still in HR livery, late in 1926, it may have gone straight to black *(Plate 334)*.

Few of them lasted long enough to receive the post-1927 style, but those which did were given a lined black livery with what seem to have been handpainted unshaded or black shaded characters, either yellow or gold, with 10in. figures — essentially Code B5 *(Plate 334)*. We can confirm only two (Nos. 14274/5) in this condition, by which time cleanliness was fast disappearing.

Plate 334 No. 14274 (ex-HR No. 94) *Strathcarron* at Fochabers in 1930, carrying lined black livery, Code B5. Note the shorter Jones type chimney with Drummond pattern top rim.

H. C. Casserley

Jones 'Loch' class 4-4-0 (LMS Nos. 14379-96; Power Class 2, later 2P)

The handsome 'Loch' class 4-4-0s were Jones' last design for the Highland Railway, and fifteen were put in service during 1896, all built by Dübs & Co. They were amongst the most powerful passenger engines in the country at the time of construction and were fitted with piston valves — the first such application in Scotland. Driving wheel diameter was 6ft. 3in. and the visual styling was very similar to the celebrated Jones 'Goods', introduced a few years earlier *(see page 203)*. Interestingly, three more were built some twenty years later, entering service in 1916 (on the order of the Railway Executive War Committee) to meet acute motive power shortages during World War I. They were built at the Queens Park Works of the North British Locomotive Company in succession to Dübs & Co.

Plates 335 (Left Below) & 336 (Above) These two views show the essential differences between the 'Lochs' as received by the LMS — No. 14387 (ex-HR No. 127) *Loch Garry* and, as rebuilt with larger boiler, No. 14391 (ex-HR No. 131) *Loch Shin*. The livery codes are A1 and B4 respectively. Note the curiously 'wrong' radius of the name *Loch Garry* in relation both to splasher and wheel centre.

Photomatic

By the time the LMS received them, the piston valves had been replaced (by Drummond) with balanced side-valves — but they were otherwise little changed *(Plate 335)*. Starting in the 1920s, many engines were rebuilt with boilers of Caledonian type. These were rather bigger and, fitted with a shorter McIntosh chimney, they gave a quite different but still well-balanced line to the engine *(Plate 336)*. With these rebuilds, the cab front spectacle glass was reduced in size and changed in shape. Finally, some of the small-boilered 'Lochs' were fitted with replacement Drummond chimneys without reboilering *(Plate 337)* one or two had the characteristic smokebox wingplates removed but retained the original Jones chimney and boiler *(Plate 338)* and some had wingplates removed and received Drummond chimneys *(Plate 339)*. We have tried, in the livery summary, to make these differentiations.

For the record, the reboilered engines were Nos. 14379-83/5-6/90-2, all between 1924 and 1928. In general, these engines outlived the others by a few years.

We believe that all the 'Loch' class received the full red livery, Code A1, and have positively confirmed all but Nos. 14385/94. These last two were in receipt of the early version of the post-1927 livery, essentially Code B5, and it may be that this was their first LMS repaint (circa 1928/9). Quite a number of others received this first lined black livery with 10in. figures (unshaded) seemingly hand-applied rather than transfer stock *(Plate 340)*.

From around 1930 onwards, 14in. Midland or 12in. standard figures, both of the countershaded style, were applied in roughly equal proportions. By now, some engines were being shopped at St. Rollox (ex-CR) and others at Kilmarnock (ex-G&SWR) and this may explain the insignia difference — but the hypothesis is a little tenuous. Late survivors received 10in. figures (yellow shaded red) frequently with plain black livery, Code C21 *(Plate 341)*. It is also worth mentioning that No. 14392 *Loch Naver* had its name rendered in a straight line, rather than in a curve, until it reached unlined black livery — *Plate 342*. In the summary *(on page 176)*, we have tried to sort out these various differences.

Plates 337 to 339 These three pictures show subtle changes within the original series of 'Lochs'. No. 14395 (ex-HR No. 71) *Loch Garve* has a replacement Drummond chimney but retains wingplates, whereas the two views of No. 14384 (ex-HR No. 124) *Loch Laggan* reveal that it lost its wingplates before the original chimney was replaced. All engines are lined black, but note that No. 14384 went from 14in. figures (Code B4) to 12in. figures (Code B3) when the chimney was changed. No. 14395 is also Code B4.

*G. Coltas,
Gavin Wilson Collection
and Photomatic*

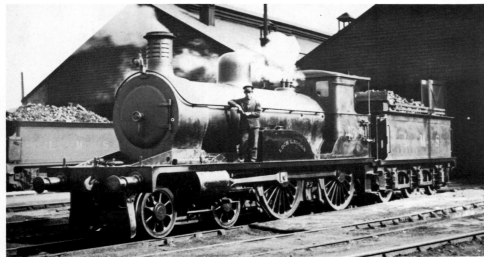

Plate 340 (Above) The lined black livery with 10in. unshaded hand-painted figures, Code B5, on re-boilered No. 14382 (ex-HR No. 122) *Loch Moy* at Inverness in May 1928.
H. C. Casserley

Plate 341 (Below) The plain black livery Code C21 on reboilered No. 14379 (ex-HR No. 119) *Loch Insh* at Aviemore in 1946.
H. C. Casserley

Plate 342 The unique *Loch Naver* with its straight name (LMS No. 14392, ex-HR No. 132) pictured at Perth in 1928, carrying livery Code A1. It was not until it received a plain black livery during the war that the name was painted in the usual fashion. Throughout its lined black phase it remained straight.

H. C. Casserley

Livery Samples

All post-1927 liveries had letters 'LMS' at 53in. centres

a) Locos in original condition — i.e. with (normally) Jones chimney, boiler and smokebox wingplates
Code A1   14382-4/6-9/93/5-6
Code B5   14385§/93-4/5§

§ *Drummond chimney*

b) Locos with small boiler without wingplates; and either Jones or Drummond chimney
Code A1   14396
Code B5   14388§
Code B3   14384§
Code B4   14384

§ *Drummond chimney*

c) Reboilered locos
Code A1   14379-81/90-2
Code B5   14379/81-2
Code B3   14380-2/5-6
Code B4   14383-5/91-2
Code C21   14379/85/92

Finally it is worth noting that two of these engines (Nos. 14380 *Loch Ness* and 14390 *Loch Fannich*) remained in pre-1928 red livery until 1937, the latter being withdrawn in this condition. Both were in reboilered form at the time.

The Drummond 'Ben' class 4-4-0s (LMS Nos. 14397-416 and 14417-22; Power Class 2, later 2P)

The 'Ben' class was Peter Drummond's first Highland design, introduced in 1898, and the first twenty, later known as the 'Small Bens' were very similar to brother Dugald's contemporary engines for the LSWR. They were not all that different, visually, from many other Drummond-designed (or inspired) types on the Caledonian and North British railways, and were the first Highland tender engines with inside cylinders. Driving wheels were 6ft. and although all, originally, had six wheel tenders, four of them (LMS Nos. 14398/409-11) were later fitted with eight wheel tenders (exchanged from 0-6-0s) with inside axleboxes. *Plates 343 & 344* show the class, more or less in original condition, with the two tender variations.

Plates 343 & 344 These views show the original 'Ben' class with its two different tender styles. No. 14404 (ex-HR No. 8) *Ben Clebrig* at Dingwall, in 1928, undergoing fire cleaning, has the original 6 wheel style — note particularly the unlined tender sideframes, probably black. This was a common but not quite universal feature of ex-HR engines when red. No. 14411 (ex-HR No. 15) *Ben Loyal* was one of the four engines of the class never rebuilt (and the last of them to be scrapped) and this circa 1926 view clearly shows the dome top safety-valves which it retained throughout. Both engines are red, Code A1.

H. C. Casserley and Stephen Collection, courtesy NRM

The building of the 'Small Bens' was shared between outside contractors and the Highland Railway itself, and was spread over an eight year period, the last coming into service in 1906. Shortly after the Grouping, the LMS (as with the 'Loch' class) began to fit Caledonian type boilers to the engines (1927-30) but this time the replacement boilers/chimneys were so similar to the originals in both appearance and size that the visual appearance hardly changed. The large support below the safety-valves revealed the newer boilers, while the safety-valves on the dome identified some of the originals. To add confusion, most of the original boilers also had their safety-valves removed to a position over the firebox but, since these did not have a base support, it was still possible to identify the two series — *Plates 345 & 346*. Moreover, the residual engines with original boilers seem to have retained their smokebox wingplates *(Plate 345)*. Nos. 14407/11/13-14 were never reboilered.

Plates 345 & 346 These views show how difficult it was to recognise a re-boilered 'Ben' from an original version which had had its safety-valves moved to the firebox top. No. 14413 (ex-HR No. 17) *Ben Alligan* has the original boiler, No. 14402 (ex-HR No. 6) *Ben Armin* has the Caledonian type. Note the massive safety-valve support and lack of wingplates. Both engines are lined black, Codes B5 and B3 respectively.

Authors' Collection and Photomatic

In 1908, two years after the last of the first twenty 'Bens' had been built, Drummond introduced a larger version of which six were built (Nos. 14417-22). These engines, again somewhat similar to LSWR designs, were known as 'Large Bens' or 'New Bens', leading in consequence to the 'Small' classification for the first series. As with the 'Small Bens', tender variations were present. The first four originally had six wheel tenders and the last two had the eight wheel type, but the first two received replacement eight wheel tenders again from 0-6-0s, thus leaving LMS Nos. 14419-20 as the only two with six wheel tenders. The last two eight wheel tenders, supplied with the engines from new, were of a larger capacity than the previous eight wheel type, a fact which is readily seen by comparing *Plates 347 & 348*. The six wheel version is seen in *Plate 349*.

Plate 347 (Above) 'Large Ben' No. 14422 (ex-HR No. 62) *Ben a' Chaoruinn* in original saturated condition with larger 8 wheel tender — *see main text*. The livery code is A1.

Stephen Collection, courtesy NRM

Plate 348 (Below) Superheated 'Large Ben' No. 14418 (ex-HR No. 63) *Ben Mheadhoin* with smaller 8 wheel tender originally fitted to an 0-6-0 — livery Code A1.

Stephen Collection, courtesy NRM

Plate 349 Superheated 'Large Ben' with 6 wheel tender, No. 14420 (ex-HR No. 65) *Ben a' Chait*. Note the lack of a Westinghouse pump compared with *Plate 348*. The livery code is A1.

Stephen Collection, courtesy NRM

These pictures also reveal the difference between the saturated and superheated state of these engines. After the Grouping, the LMS fitted all six with superheaters and extended smokeboxes between 1924 and 1927. We believe that only No. 14417 *Ben na Caillich* and No. 14422 *Ben a' Chaoruinn (Plate 347)* received LMS livery before superheating — both Code A1.

The 'Small Bens' outlived the enlarged development, ten of them reaching BR, but only three were renumbered (Nos. 14398/9/404). The first of these, *Ben Alder*, became quite a 'cause célèbre' having been selected for preservation. It languished until the mid-1960s until an act of official vandalism sealed its fate in 1966. There was, at the time, a fetish for official preservation only in 'as built' condition and No. 14398 was, of course, by then carrying a Caledonian boiler. Moreover, there was at the time no National Railway Museum either, so it was lost. The lack of a genuine Scottish locomotive (other than those transferred to the Glasgow Museum) in the 'National' display at York must, in retrospect, be considered a grave omission and *Ben Alder* would have been a good choice, especially with its CR 'overtones'.

The 'Large Bens' disappeared between 1932 and 1937, the last survivor being the last of the series built, No. 14422 *Ben a' Chaoruinn*.

The 'Bens', small and large, were fairly consistently painted in LMS days. All confirmed red examples were Code A1 and we have only failed to verify Nos. 14401 and 14408 in pre-1928 colours. They must surely have received them.

Plate 350 Superheated 'Large Ben' No. 14417 (ex-HR No. 61) *Ben na Caillach*, pictured circa 1928 in the post-1927 red livery, Code A5. The name should be spelt 'Caillich' — obviously not a Lochgorm paint job!

BR (LMR)

Plate 351 'Small Ben' No. 14398 (ex-HR No. 2) *Ben Alder* in the mid-1930s' lined black livery, running with a 'Barney' tender, Code B3.

BR (LMR)

Plate 352 'Large Ben' No. 14417 is seen here again in lined black, Code B4, during the later 1930s. The name was still spelt wrongly — *see also Plate 350*.

Photomatic

After the livery change, lined black was adopted except on No. 14417 (repainted red with 1928 characters — *Plate 350*). Some received an early form of 10in. numeral (as did the 'Lochs' and many other Highland types). Thereafter, 12in. and 14in. figures, gold with red countershading, were the usual practice, the 12in. size rather predominating on the 'Small Bens' *(Plate 351)*, but 14in. being rather more typical for the 'Large Bens' *(Plate 352)*. Long-term survivors were plain black with yellow/red insignia, and we give a fair sample *overleaf* of all liveries.

Livery Samples
All post-1928 liveries had letters at 53in. centres

a) 'Small Ben', original boilers
Code A1   14397§/402/4§/5/7/8/10/11§/2-4/6
Code B5   14398/407*/13
Code B3   14411§

NB We have only given examples where the safety-valve position is positively confirmed
   No. 14416 had no number or LMS coat of arms with this livery
   The § mark denotes those still carrying safety-valves on the dome
   *Wingplates removed and external ejector pipe fitted

b) 'Small Ben', replacement boilers
Code A1   14402/5‡/6/9/15 (almost certainly red before reboilering as well)
Code B5   14398-400/2
Code B4   14401/8/10/2/5-6
Code B3   14397-400/2-6/8-9
Code C21   14397/9/404/9/15
   ‡tender side frames lined out — see also Plate 343

c) 'Large Ben', superheated
Code A1   14417-22
Code A5   14417 (possibly also Code B5 later)
Code B4   14417-9/22
Code B3   14422

Plates 353 & 354 Alternative lettering styles on No. 14406 *Ben Slioch* and No. 14408 *Ben Hope*, both being lined black engines. The serif pattern was by far the most common. Note the different works plates.

H. C. Casserley and Authors' Collection

Cumming 4-4-0 (LMS Nos. 14522-3; Power Class 3, later 3P)
These two large engines, built in 1916 and named *Snaigow* and *Durn* respectively, were the first 4-4-0s in Britain to have outside cylinders and valve gear. They could hardly be considered to be a failure, but their inability to accept CR type replacement boilers, and the fact that there were only two of them, pretty-well condemned them to the scrapheap in 1935 and 1936, *Snaigow* being the last to go. A much better fate lay in store for Cumming's more numerous 4-6-0 types, the 'Clans' and 'Clan Goods' *(below)*.

We show opposite side views of the two 4-4-0s in *Plates 355 & 356* showing both livery styles carried by both engines. Interestingly, both of them seem to have received the lined black during the early (circa 1928-9) period, when 10in. hand-painted figures seem to have been popular on the Highland section. We have no record of either of them having received the later 12in. or 14in. figures, but it is quite possible that one or both of them did.

Plates 355 & 356 *Snaigow* and *Durn*, Nos. 14522/3 (ex-HR Nos. 73/4), in the two liveries carried at some time by both of them — Codes A1 and B5. Note again the obvious absence of tender sideframe lining on No. 14522 *(see Plate 343)*.
BR (LMR) and Authors' Collection

Drummond 'Castle' class 4-6-0 (LMS Nos. 14675-93; Power Class 3, later 3P)

This very successful design, of which another fifty, all but identical examples, were built for the French State Railways in 1911, was, in effect, Jones' last design. Details had been prepared before his resignation and Peter Drummond made but a few changes — cab shape, boiler mountings, marine type big ends, steam reverser, etc. The first sixteen (Nos. 14675-90) were also fitted with the double bogie, inside-bearing tenders of patently Drummond (LSWR?) inspiration, but in all essentials the engines were the passenger version, with 5ft. 9in. wheels, of the Jones 'Goods' type *(below)*. During LMS days, dating from the late 1930s, three of them (Nos. 14681/6/90) were given replacement six wheel tenders prior to going on the Oban (ex-CR) line. The tenders were exchanged with those from Nos. 14691-3.

These sixteen engines were build in batches over a considerable period between 1900 and 1913/14. The last four of them (Nos. 14687-90), appeared under W. M. Smith's term of office and had extended smokeboxes surmounted by chimneys with a small wind deflector or capuchon. *Plates 357 & 358* represent the two versions of the first sixteen 'Castles'.

Plates 357 & 358 No. 14677 (ex-HR No. 142) *Dunrobin Castle* and No. 14688 (ex-HR No. 27) *Thurso Castle*, both in livery Code A1, represent the 5ft. 9in. 'Castles'. Note the longer smokebox and capuchon chimney on No. 14688.

A. G. Ellis

The last group of three were ordered by Cumming (Nos. 14691-3) and came out in 1917 with enlarged driving wheels (6ft.) and a boiler with an enlarged heating surface. They were fitted with six wheel tenders and an example is shown in *Plate 359*. They had the longer smokebox of the 14687-90 series, but reverted to the more conventional chimney.

Plate 359 No. 14693 (ex-HR No. 59) *Foulis Castle*, livery Code A1, was the last of the series to be built. Note the 6 wheel tender and the bigger splashers of this 6ft. series.

BR (LMR)

Unlike the Jones and the Drummond 4-4-0s, which, as recorded above, underwent considerable rebuilding processes affecting their visual lines, the 'Castles' remained substantially 'as built' throughout the LMS period with only minor detail changes which tended to be commonplace at that time throughout the land. All were, we believe, painted red during the first phase, and we have confirmed most of them. Two at least (No. 14676 *Ballindalloch Castle* and No. 14680 *Murthly Castle*) had no Company markings at all *(Plate 360)* and another (No. 14687 *Brahan Castle*) received its first post-1927 insignia with the red livery *(Plate 361)*.

Plate 360 A red 'Castle' without an LMS crest or a tender-side number — No. 14676 (ex-HR No. 141) *Ballindalloch Castle* is pictured at Dingwall in 1928.

H. C. Casserley

Plate 361 No. 14687 (ex-HR No. 26) *Brahan Castle* received a red livery with post-1927 insignia, Code A5. The hand-painted nature of the figures is clearly seen. There is no sign of shading to the insignia, but we cannot be certain it was not there.

BR (LMR)

One or two received lined black with the early 10in. figures, but the standard post-1927 style quickly settled down, almost universally with 12in. figures, countershaded (Code B3 — *Plate 362*).

The few post-war survivors were unlined black with 10in. yellow/red figures (Code C21 — *Plate 363*) and the last survivor, No. 14690 *Dalcross Castle* with a six wheel tender, was in this style when withdrawn in 1947.

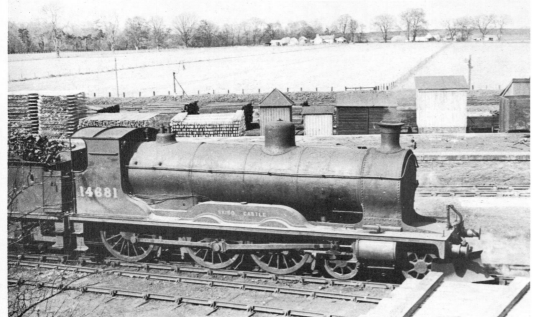

Plates 362 & 363 Lined black livery (Code B3) and plain black (Code C21) on Nos. 14686 (ex-HR No. 35) *Urquhart Castle* and 14681 (ex-HR No. 146) *Skibo Castle*, photographed circa 1936 and 1946 respectively. Note the ugly replacement dome cover.
Photomatic and H. C. Casserley

Plate 364 (Right) No. 14762 (ex-HR No. 49) *Clan Campbell* in the pre-1928 period, livery Code A1.
Stephen Collection, courtesy NRM

Livery Samples

The following list is divided into three principal groups of engines. Tender type is as described above unless annotated § (six wheel tender replacing original eight wheel type):

a) Original series (Nos. 14675-86)
Code A1   14675-7/80-2/84-6 (14676/80 originally without Company markings)
Code B3   14675-9/81§/82-5/6§ ⎫
Code B4   14684   ⎪
Code B7   14683   ⎬ letter centres at 53in.
Code C21   14678/81§   ⎪
                                ⎭

b) Second series — extended smokebox (Nos. 14687-90)
Code A1   15688-90
Code A5   14687
Code B3   14687-8/90 (14690 carried this livery with both tender styles) ⎫ letter centres 53in.
Code C21   14690§   ⎭

c) Third series — 6ft. wheels, six wheel tenders (Nos. 14691-3)
Code A1   14691-3
Code B5   14691-2 ⎫ letter centres at 53in.
Code B3   14691-2 ⎭

Cumming Clan class 4-6-0 (LMS Nos. 14762-9; Power Class 4, later 4P)
We conclude the Highland passenger tender engines with the largest and most powerful of the 4-6-0s designed for that system. They were built in two batches, four each in 1919 and 1921. They may never have been built at all (or at least to such a number) had it not been for the forced selling of Smith's 'River' class engines to the Caledonian in 1915, already considered on *page 37*.

Be that as it may, the 'Clans' were essentially a 6ft. passenger version of the already existing 5ft. 3in. goods engines designed by Cumming in 1917, just as the 'Castles' were developed from the Jones 'Goods' engines some years earlier. Interestingly, a later generation persisted in referring to the goods engines as 'Clan Goods', even though they preceded the 'Clans' into service.

The 'Clans' proper were fine modern engines with outside cylinders and valve gear, and having a visual appearance in line with the 4-4-0s *Snaigow* and *Durn (above)*. A noteworthy feature was the very long cab roof overhang — encroaching well over the tender front — which gave good protection for the crew. The engines changed very little during LMS days. They were all Westinghouse-fitted and this feature continued during the LMS 'red' period but was gradually changed as the LMS moved into the 'vacuum only' period *(Plates 364 & 365)*. Withdrawal did not commence until 1943. The last to go, No. 14767 *Clan McKinnon* achieved its BR number before scrapping in 1950.

Plate 365 This view, photographed from almost the same angle as that in *Plate 364*, shows *Clan Stewart* (No. 14765, ex-HR No. 53) without Westinghouse pump and in lined black livery, Code B3. Note the block style name — a little unusual — and also the rivetted smokebox.
Gavin Wilson Collection

An interesting 'role reversal' of these engines and the ill-starred 'Rivers' took place in LMS times. The 'Rivers' eventually came back to the Highland section; while the 'Clans' were seen in later years on the ex-Caledonian Oban line after the Stanier Class 5 had just about taken over everywhere else!

All eight engines received the first LMS red livery, Code A1 *(Plate 364)* and all were subsequently given lined black. Quite a few had the early form with 10in. figures, but 12in. seems to have become the norm fairly soon — *Plate 366*. Many of them were probably plain black with yellow/red insignia at the close of LMS days, but we have positively identified only one *(see below)*.

Livery Samples

Lined black only, all engines were red pre-1928
Code B5   14763/5/7
Code B3   14762-8   } always with letter centres at 53in.
Code B4   14767
Code C22   14767

188

4-4-0T — all types (LMS Nos. 15010-7)
The whole group of four-coupled passenger tank engines of ex-HR origin were an absolutely fascinating bunch, and we have split them into the two wheel arrangements represented, starting with the 4-4-0Ts. These were basically in two visually distinctive series, and we take them in turn.

a) The 'Crewe type' 4-4-0Ts (LMS Nos. 15010-2) These engines, dating from the Jones period, were to all intents and purposes the tank equivalent of the 'Skye bogies' *(see page 170)* but with 4ft. 9in. wheels — and how gorgeous they must have looked in the full red livery *(Plates 367 to 369)*. As stated on *page 169*, the Highland section 'adopted' the red style with almost more enthusiasm, if possible, than Derby itself, and the small tanks must have presented a wonderful sight. Rebuilt during 1881/2 from earlier 2-4-0Ts, their visual lines are fully exemplified in the illustrations. All three were red, Code A3, and one of them, No. 15010, the last survivor, saw out its time in plain black with plain gold figures *(Plate 370)*. It was scrapped in 1932.

Plates 367 & 368 These views of 4-4-0Ts Nos. 15011/2 (ex-HR Nos. 58B/50B) show two versions of the pre-1928 red livery, Code A3. Note how much further forward the number was on No. 15012. No. 15010 was similar to this.
H. C. Casserley and Authors' Collection

Plate 366 (Bottom Left) *Clan Stewart* again, still in livery Code B3, now displays its name in serifed letters.
Photomatic

Plates 369 & 370 These two views of No. 15010 were taken three years apart in 1928 and 1931 respectively. We find the front view particularly appealing. Livery codes are A3 and C13 respectively, with LMS spacing around 40in. centres.
H. C. Casserley

Plate 371 'Yankee' tank No. 15013 (ex-HR No. 101) in the first LMS livery applied to Nos. 15013-4/6-7, Code A3.
A. G. Ellis

b) The 'Yankee' tanks (LMS Nos. 15013-17) The nickname 'Yankee' was bestowed on these engines because, in 1892, two had been built by Dübs & Co. for the Uruguay Eastern Railway, which in the end were purchased by the HR. The fact that Uruguay is not even in the same hemisphere as the real 'Yankee' territory (the USA) seems not to have been considered! The Highland obviously like them because in 1893, three more (Nos. 15015-7) were added. No. 15015 was never renumbered by the LMS and No. 15017 is the only Highland engine to have totally eluded our own searches through the archives.

As far as we know, the LMS painted all, except No. 15015, red *(Plate 371)* and, by the time they were received, No. 15014 had received a Drummond chimney and slightly enlarged boiler *(Plate 372)*. This engine is the only one we have seen photographed in the post-1927 livery and we believe it to have been lined out *(Plate 373)*. It is possible that No. 15013 may have been similar (both scrapped 1934), but the others had all been withdrawn before the livery change.

Plates 372 & 373 4-4-0T No. 15014 (ex-HR No. 102) had a Drummond chimney and boiler, as seen in these two views showing its pre and post-1928 livery styles, Codes A3 and B2 respectively. Under a magnifying glass, the red lining and countershaded insignia are just discernible.

H. C. Casserley and Authors' Collection

0-4-4T-all types (LMS Nos. 15050-4; later Power Class 0P)
As with the 4-4-0Ts, this second set of four-coupled tanks consisted of two different types. No. 15050 was a Jones 'one-off' from 1890, and the others were of Drummond design, dating from 1905.

Dealing first with No. 15050, this engine began in 1890 as an 0-4-4ST, built by Jones for the Strathpeffer branch and bore the name of that town. It was unusual both in carrying a saddle tank and in that it was the only Jones design for the HR with inside cylinders. As received by the LMS, it was as rebuilt with side tanks and Drummond boiler in 1901. Renamed *Lybster* in 1903, it went to work from Wick. It remained in that area until withdrawn (1929) but had lost its name before the LMS period. We trust we shall be forgiven for offering two views of this charming little engine, resplendent in LMS colours, in *Plates 374 & 375*.

Plates 374 & 375 Two evocative views of the pretty little 0-4-4T, No. 15050 (ex-HR No. 53A), at Wick in 1928 — livery Code A3.
H. C. Casserley and A. G. Ellis

No. 15050 might have been an oddity but Nos. 15051-4 were almost incredible. They were built to Drummond's design in 1905 for branch working and were the last engines to be constructed at the Highland's own works at Lochgorm. Perhaps this was why, apart from an early withdrawal in 1930 (No. 15052) nobody had the heart to scrap them! Two survived into BR days and eventually became the last two ex-HR engines in service, the last to go being No. 55053 in 1957.

Of exceptionally neat design, their appearance hardly changed in over fifty years and, had they lasted a little longer, we feel sure that one of them would, by now, be preserved with the Strathspey Railway! They were always photographer's favourites and we present a potted pictorial history in *Plates 376 to 381*, again presenting our apologies (if necessary) for slightly 'over-egging' the pudding!

Plates 376 & 377 — captions overleaf

Plates 376 to 379 All four Drummond 0-4-4Ts are illustrated here to show the progression of LMS styles from the 1920s to nationalisation. Nos. 15051/2 (ex-HR Nos. 25/40) are red, Code A3; No. 15053 (ex-HR No. 45) is lined black, Code B3 and No. 15054 (ex-HR No. 46) is plain black, Code C21. Note also the subtle detail changes as time progressed — e.g. safety-valves, mudguards, smokebox rivets, etc.

A. G. Ellis, BR (LMR) and Authors' Collection

Plates 380 & 381 (Right) No. 55053 operated between The Mound and Dornoch in early BR days and is portrayed both in early BR plain black, in 1949, and the later fully-lined livery of 1955, specially applied during its last full 'shopping'.

H. C. Casserley and Gavin Wilson Collection

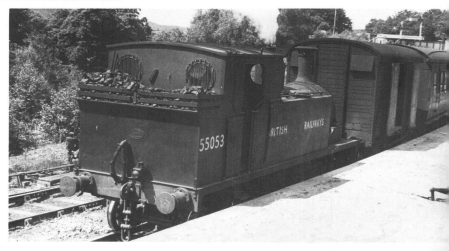

Drummond 0-6-4T (LMS No. 15300-7; Power Class 4, later 4P)
The last ex-HR passenger engines to be considered are the eight imposing 0-6-4Ts, introduced by Drummond in 1909 and built in two batches of four each notionally in 1909 and 1911 but whose delivery stretched into 1912. They were the last engines designed by Drummond for the Highland and were clearly the basis of his later design 0-6-2T for the G&SWR already considered *(page 147)*.

The engines were designed for banking duties, especially between Blair Atholl and Druimuachdar, but were also used on local passenger and goods trains. We feel that, as with the GSWR derivative, the LMS could equally well have classed these engines as freight types, since they had but 5ft. driving wheels; but their 'passenger' classification was all the excuse needed to array them in the full panoply of lined crimson lake during 1923-7.

Plates 382 & 383 Opposite side views of 0-6-4Ts No. 15302 (ex-HR No. 65, later No. 68) and No. 15300 (ex-HR No. 39) in red and lined black styles of painting, Codes A1 and B4. Note the change in safety-valves on No. 15300 as a result of reboilering. Nos. 15306/7 were similarly altered.
A. G. Ellis

During the LMS period they changed but little and became lined black in 1928. However, one picture exists which suggests that No. 15302 may have remained red after the insignia change *(Plate 384)*. The two characteristic liveries were with 18in. figures during the red phase *(Plate 382)* and 14in. Midland figures in the lined black period *(Plate 383)*. Confirmed examples of the two are as follows:

Red, Code A1, 15300-3/5/7
Lined Black, Code B4, 15300-1/4-6, all 53in. letter centre spacing

We think it likely that the omitted examples mostly conformed to this pattern. Withdrawal was completed between 1932 and 1936, the last to go being No. 15300 at the end of the year. It is just possible that No. 15303 (withdrawn 1932) remained red throughout.

Plate 384 This rear view of No. 15302 in post-1927 style insignia strongly suggests a red livery when examined closely — Code A5. The view was taken at Forres in 1930.

H. C. Casserley

Freight Classes

As with passenger engines, so too with the freight types, tender engines predominated on the HR, so we have not split the two categories. In point of fact, as we mentioned on *page 169*, the Highland had a much larger proportion of engines classed by the LMS as 'passenger' than it had freight engines. The latter numbered not much more than one quarter of that part of the fleet which received LMS numbers (41 out of 150 to be precise).

The reason is not hard to find. The Highland was a hilly railway and most of its passenger engines had quite small driving wheels — nothing bigger than 6ft. 3in., and mostly smaller. In consequence, many of the so-called passenger types were not unsuited to goods working, and the Highland had evolved a 'mixed traffic' concept of locomotive operating long before the phrase had been invented. It is small wonder that the Highland district took to the Stanier Class 5 4-6-0 like 'ducks to water'. In many respects what they were being offered was a more modern and efficient version of that to which they had become accustomed since the 1890s (and earlier).

Nevertheless, the Highland did have a group of engines predominantly of the freight category, and we finish this part of our survey by looking at them in LMS number order. Needless to say, quite a lot of them were used on passenger or mixed trains.

0-6-0T — all types (LMS Nos. 16118-9; 16383; 16380-2)
Two basic classes were included in this group, three of each type, and they were the only freight tanks owned by the Highland Railway.

a) Stroudley 0-6-0T (LMS Nos. 16118-9; 16383) These three engines were the HR forerunners of Stroudley's more famous LB&SCR 'Terrier' tanks but, in actual fact, only one of them (LMS No. 16118) originally *Balnain* and later (1902) *Dornoch*, had actually been built during Stroudley's term of office. No. 16383 was wrongly renumbered in 1923 (more precisely, wrongly identified!) and should, strictly speaking, have been No. 16120. This was never rectified. The general characteristics of the design are clear from *Plates 385 & 386* which indicated clearly that two tank lengths were present by LMS times. The third member of the class (No. 16119) was more like No. 16383 and both of them were post-Stroudley in date of build.

Plates 385 & 386 LMS 0-6-0Ts Nos. 16118 (ex-HR No. 56B) and 16383 (ex-HR No. 49A), both in pre-1928 livery, Code C4, were photographed at Inverness in the 1920s. The physical differences between the genuine Stroudley engine, now rebuilt with larger tanks (No. 16118) and the later-built example are readily apparent. Note also the different insignia placement.

Stephen Collection, coutresy NRM

The LMS painted them black, but No. 16119 never ran in pre-1928 livery, still being in HR colours in May 1928. However, it was the only one to be painted in the post-1927 style, at which time it was given red lining *(Plate 387)*. This was a not unusual Highland district variation in the early days of the 1928 livery, and was even applied to LMS standard engines from time to time *(see Volume One, introductory plate to Part 1)*. We do not know at what time officialdom stepped in to force conformity, but we have confirmed at least one example of all ex-HR freight designs in lined black circa 1928-30, except for the 0-6-0s, and these are considered below.

No. 16119 was the last survivor of the Stroudley type and went to scrap in 1932.

Plate 387 0-6-0T, No. 16119 (ex-HR No. 57B) is seen in post-1927 livery, Code B5, with letters at 40in. spacing. This was one of quite a number of ex-HR freight types given lining after the livery change. The lining itself is not easy to see, but is most clearly indicated on the bunker side.

Authors' Collection

b) Drummond 0-6-0T (LMS Nos. 16380-2; Power Class 2, later 2F) These three solid-looking outside-cylindered 0-6-0Ts were built at Lochgorm in 1903 for shunting duties. The wheels, cylinders and motion came from dismantled 2-4-0 goods engines and the boilers were second-hand from rebuilt 2-4-0 passenger types. Hardly surprisingly, they were nicknamed 'Scrap' tanks! Nevertheless, they had almost thirty years of life, not vanishing until 1932 with the withdrawal of No. 16381.

The LMS liveries are clear from the pictures of Nos. 16382 (pre-1928) and 16381 (post-1927) — *Plates 388 & 389*. Once again, lining was applied with the post-1927 style. We can also confirm No. 16381 in the identical style to No. 16382, illustrated, and No. 16380 in post-1927 style with insignia arranged like that on No. 16381, illustrated. We cannot say whether it, also, was lined out.

Plates 388 (Left) & 389 (Top Right) Opposite side views of 'Scrap' tanks Nos. 16382 (ex-HR No. 24) and 16381 (ex-HR No. 23) show the pre and post-1928 painting styles adopted for these engines. Once again, the post-1928 version is lined out. The livery codes are C4 and B5 respectively.

A. G. Ellis and H. C. Casserley

Drummond 0-6-0 (LMS Nos. 17693-704; Power Class 3, later 3F)
These engines, built more or less contemporaneously with the original 'Small Bens' were always known as the 'Barneys'. The twelve examples came out in three batches, all originally slightly different visually, viz:

Nos. 17693-8	Original style, fitted with eight wheel tenders, built 1900
Nos. 17699-702	Water tube fireboxes, six wheel tenders, built 1902
Nos. 17703-4	Original arrangement, six wheel tenders, built 1907

At a later stage, the six bogie tenders were exchanged with the 'Bens' as explained on *page 179*. Thus the LMS received them all with six wheel tenders *(Plate 390)*. By this time, one of the water tube firebox engines (No. 17701) was in the process of being rebuilt with a conventional boiler, and No. 17699 quickly followed in 1924 and No. 17700 in 1925. Thus, only one engine ran for any length of time in LMS ownership with the water tube type boiler (No. 17702 — *Plate 391*). This engine was not modified until 1934.

Plate 390 Drummond 0-6-0 No. 17694 (ex-HR No. 135), livery Code C1, represents the 'as received' condition of most of the ex-HR 'Barneys' in LMS days. Note the LMS panel placed somewhat higher than the mid point of the cab side — a common feature on this type with this livery.
Authors' Collection

Plate 391 No. 17702 (ex-HR No. 21) was the only Highland 0-6-0 to run for any time in LMS days with its water tube firebox and the apparatus is clearly visible. This picture dates from circa 1930 and shows livery Code C13.

Authors' Collection

In the meantime, and as with the 'Small Bens', the 0-6-0s began to receive Caledonian type boilers. The replacement boiler was the same as that used for the 'Bens' and visual identification of the reboilered engines was not particularly easy. The much more substantial support for the firebox-mounted safety-valves was the best indication of the new type *(see Plates 392 to 395)*. One or two engines sported dome-mounted safety-valves before reboilering and, astonishingly, one of the old type boilers was put back on No. 17695 in 1943 *(Plate 395)* after it had run for twelve years with a Caledonian boiler. This it carried until 1948.

Plate 392 A few 'Barneys' with original boilers retained safety-valves on the dome. No. 17699 (ex-HR No. 18) was one which was late to be reboilered, if at all *(see page 202)* and carries plain black livery with 14in. Midland numerals — Code C14.

Authors' Collection

Plates 393 & 394 Opposite side views of two reboilered 'Barneys' displaying the two most common insignia variations of the post-1927 period. No. 17704 (ex-HR No. 55, later 37) has 14in. figures, Code C15, and No. 17693 has 12in., Code 14. The pictures were taken at Inverness in 1936 and 1937 respectively. Note the safety-valve support — the best visual indication of the new boiler — which is of a slightly larger pattern on No. 17693. The decorative smokebox 'star' is also interesting on this engine — a not uncommon Highland as well as Caledonian embellisment.

Photomatic

Plate 395 (Left) 'Barney' 0-6-0 No. 17695 (ex-HR No. 136) is seen in wartime livery, Code C21, and is carrying the replacement original type boiler fitted from 1943 to 1948. Note the dome-mounted safety-valves.

H.C. Casserley

Plate 396 A reboilered 'Barney' with stovepipe chimney (fitted circa 1946) seen running as BR No. 57697 (ex-HR No. 138) in early BR days, circa 1949.
Authors' Collection

The reboilering of the 'Barneys' was somewhat more protracted than with the 'Bens', taking place between 1926 (No. 17693) and 1945 (No. 17698). Two engines were never given Caledonian type boilers and both were from the original water tube series (Nos. 17699/701), although there is some confusion as to whether or not No. 17699 was reboilered in 1943 *(Plate 392)*. Finally, Nos. 17697/700 received stovepipe chimneys in 1946 *(Plate 396)*. We have tried to arrange our livery summaries to take into account these differences.

Seven engines went into BR stock, three being renumbered (Nos. 57695/7-8) and the last withdrawal was No. 57695 in 1952.

The 'Barneys' were always plain black in LMS days as far as we know. We have discovered no positive examples of lining being applied in the 1928-30 period but, given the independence of thought in the district at this time, there may have been an odd maverick. We have not been able to confirm all engines in genuine pre-1928 liveries, but those that we have checked suggest that the red 'LMS' panel was generally set higher on the cabside than with most types. This was to clear the tablet exchange apparatus. We do not believe that any reboilered engines received pre-1928 livery.

After the insignia change, there were a few repaints with the small 10in. numerals as was common on the Highland at that time. Thereafter, 12in. and 14in. figures were used, the former being slightly less common, we think. Plain gold insignia were normal until the late 1930s when yellow/red became customary and with this final LMS style, 10in. figures were much more common. At all times, 53in. letter spacing was adopted.

Livery Samples

a) Original boiler
Code C1   17694/6/9§/700§/2§/3-4 (17694 had panel in *centre* of cabside)
Code C13   17696§/8§/701-2§/3
Code C14   17696
Code C15   17696§/9§
Code C21   17695§

NB 17702 still with water tube firebox on this list
  § safety-valves on dome, remainder on firebox

b) Replacement boiler
Code C14   17693/4/7/702-3
Code C14   17693/4-5/7/700/4
Code C21   17694/8/702/4
Code C22   17693

Jones 'Goods' 4-6-0 (LMS Nos. 17916-30; Power Class 4, later 4F)

The immortal Jones 'Goods' was Britain's first ever 4-6-0 when introduced in 1894, and is probably the most famous of all the celebrated Highland types. Known originally as the 'Big Goods', the fifteen engines were ordered as a single batch and all were delivered in 1894. To say that they were something of a sensation is to put it mildly and, for more than forty years, they pounded the gradients of the Highland system at the head of heavy trains. Thankfully, the first example was set aside for preservation by the LMS in 1934, and reposes now at the Glasgow Museum of Transport — arguably the most significant purely Scottish design saved for posterity.

The LMS was in no great hurry to scrap them and it was not until the Stanier Class 5 came on to the scene in quantity that serious inroads were made into the class. They just lasted into the World War II period, the last example (No. 17925) being scrapped in February 1940.

Little change was made to them during their life — there was no real need to do so — and the only superficial change during LMS days was the replacement of the louvred Jones chimney with the Drummond type (and the removal of smokebox wingplates) on many members of the class between 1925 and 1936/7. Even so, five managed to survive to scrapping without even these minor changes. These were Nos. 17918-9/21-2/8, all scrapped between 1929 and 1934. One of the others, the pioneer engine itself, received a shorter version of the characteristic Jones chimney just before the Grouping *(Plate 397)*. The more typical 'as received' condition is represented by *Plate 398* and the final appearance is typified by *Plate 399*.

Plate 397 The pioneer 'Jones Goods' (HR No. 103) pictured at Perth in 1928, as received by the LMS and carrying its new number, 17916, in the first painting style, Code C1. Note the 'short' Jones chimney.

H. C. Casserley

Plate 398 No. 17922 (ex-HR No. 109) was photographed at Perth a week earlier than No. 17916 *(Plate 397)*. It shows the same livery but features the taller chimney, normal for this class at the Grouping.

H. C. Casserley

Plate 399 No. 17920 (ex-HR No. 107) was noteworthy on two counts. It received the early LMS livery, Code C1, when carrying its Jones chimney, and was repainted the same way after removal of the wingplates and the fitting of a replacement chimney as seen here. Secondly, it retained this paint scheme until withdrawn late in 1937.

Authors' Collection

In livery terms, all repainted engines were very consistent from 1923-7, and we can positively confirm twelve of them, the other three probably being similar. At the livery change, there was the usual short-lived Highland district interval of 10in. cabside numbers and, as with most freight types, a few Jones 'Goods' (at least four) were given lining *(Plate 400)*. Thereafter, 14in. Midland figures on plain black was the most favoured of the choices available (not exclusively, however) and letter spacing was at 53in. centres — *Plate 401*. In passing, we feel we should state that some sources (unconfirmed) allege that some, at least, of the lined-out freight engines in the Highland district had yellow (straw) lining during this interim (1928-30) period. It is at least a possibility, since the lining is very clearly evident on pictures and red lining was often hard for cameras to 'see' in those days. We would welcome more information on this point. We merely list below the known lined examples, and are assuming they were lined in red, until proved to the contrary.

Plate 400 LMS No. 17928 (ex-HR No. 115) is one of at least four 'Jones Goods' known to have received a lined black 'intermediate' livery, Code B5, just after the 1928 style change. This is clearly seen displayed on a beautifully turned-out engine at Inverness in 1930. Soon afterwards, cleanliness tended to vanish from the LMS locomotive scene in the Highlands — as elsewhere!

H. C. Casserley

Plate 401 The more typical LMS livery for the 'Jones Goods' in the 1930s is seen on No. 17925 (ex-HR No. 112), Code C15. The engine now has a Drummond style chimney and lacks wingplates.

Photomatic

Livery Samples

a) Original condition, Jones Chimney/wingplates
Code C1   17916-8/20-2/4-5/7-30 (17916 with short chimney)
Code B5   17919/26/8/30
Code C13   17917-8

b) Later condition, wingplates removed, Drummond Chimney
Code C1   17920 (withdrawn 1937 in this style)
Code C15   17916-7/23-5/7/9-30
Code C14   17926

Cumming 4-6-0 (LMS Nos. 17950-7; Power Class 4, later 4F)
The last HR engines to be considered are the eight superheated goods 4-6-0s with 5ft. 3in. wheels, introduced in 1917 and built, four each, in 1918 and 1919. These were extremely powerful engines for their total weight, delivering a tractive effort of over 25,000lb. for a gross weight well under 60 tons. They did not have a class name as such, but after the 6ft. passenger version (named after 'Clans') was introduced, the 5ft. 3in. goods engines were almost always known as 'Clan Goods'.

As with the 'Clans', these engines changed very little in appearance. They had the same overhanging cab roof and, in general, lasted a year or two longer than the 'Clans' themselves. All but two (Nos. 17952/7) came to BR, and of the other six, only No. 17953 failed to receive its 5XXXX series number. The last to go was No. 57954, in late 1952.

The LMS liveries adopted closely followed those of the Jones 'Goods'. We have confirmed five in correct pre-1928 livery, and guess the others were similar. However, two of these repaints received the later round-cornered cab panel — rare on the ex-Highland engines — which suggests that repainting may have been a little slow. There were at least two lined black repaints, circa 1928-30, after which the engines generally displayed 14in. Midland figures on the cab. In the final LMS period, yellow insignia, shaded red were in use. Letter spacing was at 53in. centres throughout. *Plates 402 to 407* give a good cross-section of their LMS history and the livery summary appears on *page 208*.

Plates 402 & 403 These views show opposite sides of 'Clan Goods' engines Nos. 17950 (ex-HR No. 75) and 17952 (ex-HR No. 77) respectively. Both carry the first LMS livery but No. 17952 has the round-cornered cab panel — Codes C1 and C2.
Authors' and Gavin Wilson Collection

Plate 404 (Left) No. 17950 is seen again, at Inverness in 1928, now wearing lined black livery albeit hardly visible, Code B5. Note that, as with No. 17928 *(Plate 400)* it is still commendably clean.

H. C. Casserley

Plates 405 (Right) & 406 (Below) These views of Nos. 17957 (ex-HR No. 82) and 17954 (ex-HR No. 79) show further useful detail of these long-lived engines which might help would-be modellers. The pictures were taken at Inverness and at the Kyle of Lochalsh in 1931 and 1937 respectively, and both show standard livery, Code C15. On the original, the black shading to the LMS is clearly visible on the tender of No. 17957. Note the smokebox wheel on No. 17954 and the front coupling links fitted by the LMS well after the Grouping — *see other pictures*.

G. Coltas and A. G. Ellis

Plate 407 Our final view in this review of pre-group classes shows Cumming 4-6-0 No. 17950 yet again, this time in final LMS style, livery Code C21. It is still pretty clean possibly because it is operating a special saloon which is just visible on the left. For the record, this vehicle was an extensive rebuild (by the Caledonian and LMS) of an erstwhile West Coast dining car and it still exists — preserved by Steamtown, Carnforth and now part of the private 'Royal Scotsman' tour train.

Authors' Collection

Livery Samples

Code C1	17950-1/4
Code C2	17952/5
Code B5	17950-1
Code C15	17950-1/3-5/7
Code C14	17956
Code C21	17950-1/4/6

Chapter 8
The 'Drummond' factor on the LMS in Scotland

Having, in the last 200 or more pages, reviewed the whole of the Scottish pre-group locomotive contribution to the LMS, it seemed to us that there was still something missing in terms of rounding out the story. It is, in some ways, not too difficult for us, even allowing for our own admitted lack of expertise in several of the more detailed areas, to look at matters from the standpoint of 30 years or more of accumulated knowledge and study; yet, viewed from the perspective of the 1980s, it may be harder than we had, perhaps, reckoned for more modern day enthusiasts to realise just how uniquely different was the Northern Division of the LMS, compared with the English part of the system. This is especially true if taken against the background of the modern BR network which, for thousands of newer entrants to the field of railway history, is the only current 'scene' they have ever known. We have, therefore, taken advantage of the further 'split' in our survey *(see page vii)* to try and 'flesh-out' the story a little more than might otherwise have been possible and, after some reflection, we finally concluded that the principal contribution to the significant difference in the Northern Division could, in a very real sense, best be summed up by the one word 'Drummond'.

There are always dangers lying in wait for those who would attempt to make simple generalisations in the interests of making a complex story more comprehensible, but in the case of the Scottish engines of the LMS, we felt that the Drummond influence was so profound as to make the attempt worthwhile — and on re-reading the preceding chapters, we were made conscious of this fact to a degree we had not previously appreciated. We realised, in fact, that while we had, frequently, mentioned the Drummonds, both Dugald and Peter, in the context of specific classes, we had not done much in the way of cross-connecting the designs they produced for different systems, or, indeed said very much about either the preliminary influences upon them or their legacy to subsequent designers. This we propose to assay in these final pages.

The truly pivotal figure was Dugald Drummond on the Caledonian Railway; but even this great man did not start with a blank sheet of paper as it were. One has only to look at his early designs to realise how heavily he was, at first, influenced by one of his early chiefs, William Stroudley, later of Brighton Line fame. The original Caledonian 'Jumbos' — *see Chapter 4* — might well have been Stroudley engines in terms of their physical characteristics, having a great deal in common with the LB&SCR 0-6-0s built by Stroudley. Moreover, Stroudley himself had 'cut his teeth' on the Highland Railway and this too, was not without its subsequent influence on the LMS scene in later years. In fact, we did contemplate heading this chapter the 'Stroudley' factor at one point! However, as far as LMS engines were concerned, it was Drummond's continuation of Stroudley ideas which, in the event, was to prove rather more significant than David Jones'

Plate 408 Ex-Highland Railway 0-6-0T No. 16119 was a small, but tangible reminder of the initial Stroudley influence on the LMS locomotives of the Northern Division, although in fact built by David Jones to his predecessor's design. Its LMS livery is the same as that shown in *Plate 387*, although the lining is hardly discernible.

Photomatic

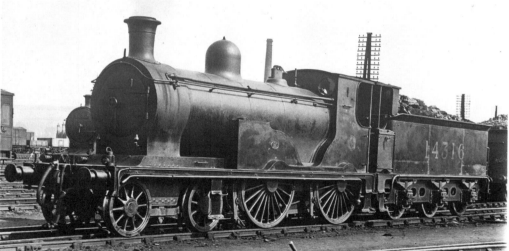

Plates 409 to 411 Undiluted Dugald Drummond — well, almost. In fact, two of these three engines were not built by him at all, but by one of his Caledonian successors (McIntosh) in the case of 'Dunalastair I' No. 14316 (ex-CR No. 726) and by brother Peter in the form of ex-HR 'Small Ben' No. 14411 *Ben Loyal* (ex-HR No. 15). The genuine article is represented by 'Coast' bogie No. 14114 — obviously photographed on the same occasion as *Plate 11*. Note, particularly, the 'Ben's' 8-wheel tender, copied from Dugald's well-nigh identical LSWR Class C8 4-4-0. LMS livery codes are A1, A1 and B3 respectively.

Authors' Collection and Photomatic

alternative development of Stroudley's methods on the Highland Railway, important though this was. Jones deviated rather more than did the older Drummond from the original Stroudley concepts, particularly in his use of outside cylinders; yet even he did not get so far away from the 'mainstream' as to render all his engines incapable of receiving 'Drummond inspired' modifications during the LMS period — the 'Loch' class 4-4-0s, for example. It was, however, left to Jones' successor, the younger Drummond, Peter, to alter the Highland Railway's locomotive details and lineaments to something whose features, recognisably, had a great deal in common with the products of brother Dugald at St. Rollox and, later, at Eastleigh (LSWR).

Peter Drummond was, of course, one of the more enigmatic figures of railway history and we have, several times, alluded to the way in which he seemed to live and work 'in the shadow of' his older brother. Take, for example, the 'Small Ben' 4-4-0s of the Highland. Visually, these were almost a carbon copy of the early Dugald Drummond type 4-4-0s for the Caledonian, as a quick comparison of *Plates 410 & 411* will reveal; moreover, these engines were, in all essentials,

Plates 412 & 413 . . . and McIntosh carries on. Saturated 'Dunalastair II' No. 14329 (ex-CR No. 773) and superheated 'Dunalastair IV' No. 14454 (ex-CR No. 47) both reveal their common ancestry which even the Pickersgill tender and later LMS boiler fittings, in the case of No. 14454, cannot seriously affect. LMS livery codes are A1 and C21 respectively.

Authors' Collection

Plate 414 The so-called Peter Drummond HR 'Castle' class 4-6-0 was, in fact, a Jones design to which Peter had added a few more typical Drummond features before it was introduced. Visually it was a nice bringing together of the Jones and Drummond developments of the original Stroudley ideas and the design was a considerable success. The featured example is No. 14675 *Taymouth Castle* (ex-HR No. 140), the first to be built, in LMS lined black livery, Code B3.

Photomatic

Plate 415 The Higland 0-6-0 'Barney', by Peter Drummond, was another almost 'carbon copy' design of his brother's work. No. 17704 (ex-HR No. 55), the last example built, is seen here in 1936 with replacement Caledonian type boiler, carrying LMS livery Code C15.

Photomatic

every bit as good as the 'Caley' breed. Likewise, too, the Highland 'Barney' 0-6-0 was not all that different from the Dugald Drummond 'Jumbo' of the Caledonian, save that when Peter Drummond was building his examples, he was, with one batch, to be influenced by Dugald's London & South Western experiments with 'water tube' fireboxes *(see page 199)*. Viewed in retrospect, this doesn't seem to have been particularly wise. Neither, for that matter, does Peter's initial work on the Glasgow & South Western — *see Chapters 5 & 6*. When Peter Drummond got to this railway, Dugald was happily pursuing some of his more eccentric ideas on the LSWR and Peter seems to have been so mesmerised by the reputation of his older brother that he seems to have been impelled to follow the 'family line', as it were — and this did the GSWR no favours at all, as we have seen. It was only in his final days on the GSWR that Peter Drummond seems to have appreciated that Dugald was not infallible and by then it was, sadly, too late for any significant forward moves to be made; for Peter died in office. Nevertheless, it is only fair to conclude this résumé of the younger Drummond by recapitulating the real achievements which stand to his credit, notably the 'Bens' and 'Barneys' of the HR, the six-coupled tank engines of the HR and GSWR and the superheated 4-4-0s and 2-6-0s of the GSWR. These engines all lasted well into LMS days — some of them beyond — and were fully in the best 'Drummond' tradition.

All of which brings us back to that colossus of the railway scene at the turn of the 19th and 20th centuries, Dugald Drummond. His presence lurked like a shadow upon the Scottish, nay even the British scene; and his immediate contemporary influence was, conceivably, matched only by Churchward of the GWR a few years later — yet he left Scotland as early as 1891; so why do we claim that his legacy was so profound on the Northern Division of the LMS? In a nutshell it is because we consider that it was when he was at St. Rollox (CR) that he did his most significant and long lasting work. Before he left the Caledonian, he had set this great railway upon a line of locomotive evolution upon which his successors could build, provided — and it is an important proviso — that they did not depart, too radically, from the fundamentally sound precepts which Drummond had established during his, relatively short, Scottish term of office on the Caledonian Railway. It is to the everlasting credit of Lambie, McIntosh and, to some lesser extent, even Pickersgill, that they did not depart too often down the interesting, but non-productive by-ways, into some of which Dugald found himself getting diverted when he moved to the LSWR. When they did (c.f. Pickersgill at times on the CR and, of course, brother Peter on the GSWR), things did tend to get a bit out of hand from time to time!

Plates 416 & 417 Peter Drummond's later 4-4-0s for the HR and the GSWR had more in common, visually, with Dugald's contemporary LSWR designs than they had with their purely Scottish ancestors. Even though they were superheated, they did not last anything like as long as the contemporary McIntosh and Pickersgill 4-4-0s of the Caledonian which had followed more closely the 'pure' line of development. Pictured here are 'Large Ben' No. 14418 *Ben Mheadhoin* (Ex-HR No. 63) and 137 class No. 14520 (Ex-GSWR No. 329), withdrawn in 1932 and 1935 respectively, LMS livery Codes A1 and B4. *Gavin Wilson Collection and Photomatic*

Dugald Drummond was at St. Rollox for only some nine years, yet it cannot be gainsaid that his influence lived on until almost the close of the LMS era as a result of the work of most of his subsequent disciples — and in this context, some of brother Peter's work can, perhaps, be considered rather maverick. For our part, we prefer to liken the Scottish work of the older Drummond as being not altogether dissimilar to that of William Stanier in a later day. Like Dugald Drummond, William Stanier was, effectively, at the head of his company's affairs for rather less time than is often realised — ten years for all practical purposes — yet his influence was to last for a far longer time than his mere period of office taken in absolute terms. Who, for example, could seriously deny that the work of Fairburn and Ivatt in the 1940s on the LMS, or Riddles and Bond after nationalisation in the 1950s for that matter, was not, in all basic essentials, a continuation of the Stanier philosophy, established during the 1930s? In this broad sense, Drummond's work on the Caledonian lasted well beyond his term of office — a fact for which the LMS had, more than once, cause to be grateful. If nothing else, it gave the means whereby a sort of 'second level' standardisation could be established in Scotland — and this cannot have been anything but beneficial in sheer costing terms before the eventual supersession of 'native' Scottish types by the chosen 'standardised' successors. Of course, given LMS thinking, it meant the fairly speedy slaughter of anything which did not fall within what might be called the 'Drummond-based' parameters, but then — and not for the first time — it is rather important that we should emphasise that the LMS Railway was not conducting its business to delight the latter day enthusiast, but was trying to work its traffic effectively! To this end, the Northern Division, in the shape of its 'Drummond Locomotive Legacy', played a not inconsiderable part, deserving, we think, rather more acknowledgement than has, hitherto, been given.

That the above paragraphs contain something of a subjective assessment of the story cannot be denied — nor would we wish to do so; but one day, we hope, some more highly qualified author(s) than ourselves will try to write the definitive book tracing the influence in Britain as a whole of the basic Stroudley-cum-Drummond approach to locomotive matters in the pre-long travel, long lap valve era, not to mention giving a more authoritative account of the pre-Churchward influence on things. Meantime, we have contented ourselves by finishing this part of our work with a pictorial reminder of such parts of that story which characterised the LMS in Scotland. If nothing else, it has, at least, given us a heaven-sent opportunity to include a few more favourite pictures of engines which could not find a place in the more specialised chapters — not to mention maintaining the goodwill of our Scottish friends!

Plates 418 & 419 The continuing Drummond influence on the LMS in Scotland is well exemplified by these two engines, both photographed in the 1930s. 0-4-4T No. 15267 was built by the LMS to Pickersgill's variation of the McIntosh design — *see page 55*; yet its visual lines are virtually pure Drummond and very reminiscent of Dugald's M7 class 0-4-4Ts for the LSWR. The 0-6-2T, No. 16907 (Ex-GSWR No. 8) was Whitelegg's variant of Peter Drummond's design — *see page 147*. This, in turn had been developed from the earlier Peter Drummond 0-6-4T for the HR and the picture, interestingly, was taken on the Highland section at Blair Atholl in 1937. LMS liveries are Code B4 and C15, although the 0-4-4T may, possibly, have been red — *see also Plate 109*. This view dates from July 1933 but the tonal difference between boiler and smokebox is still present.

Photomatic

Plate 420 We conclude our survey with a symbolic picture, taken in early BR days at Corkerhill in 1948, which shows the three dominant influences on the LMS in Scotland — Drummond, Stanier and Midland — in the shape of HR 'Barney' No. 17693 (Ex-HR No. 134, livery Code C22), rebuilt 'Royal Scot' No. 6103 *Royal Scots Fusilier* (livery Code B12) and an unidentified LMS standard Class 4P compound 4-4-0 in the 9xx series. It seems a highly appropriate way by which to link this book with *Volume Four*.

H. C. Casserley

Appendix

Corrections and Additions to Volume One and Two

Corrections
At the time of going to press, only a few further corrections had come to light which affect the meaning of the text — disregarding obvious misprints, of course. These are as follows:

Volume One
Page 112	Locomotive Superintendents List — references to 1890 should read 1891 throughout.

Volume Two
Page vi, lower picture	LYR No. was 49 and caption should read 'feed water pump and heater'.
Page 41, para 3, line 4	For 'Plate 93' read 'Plate 88'.
Page 45, subheading	LMS Nos. '6515–6004' should read '6515–6600'.
Page 66, para 4	First word should be 'They' not 'There'.
Page 78, para 2, line 1	'Coal Tanks' should read 'Coal engines'.
Page 108, 0-8-0 numbering	Class D LMS Nos. were 9002–64.
	Class C1 LMS Nos. were 8968–9001.
Page 113, line 6	'Nos. 8968/70-1/4/7/96/' should read 'Nos. 8968/70-1/84/7/96'.
Page 173, para 5, line 1	'four' should read 'few'.
Page 181, last words	should read 'not infrequent'.
Page 212, para 3, line 1	should commence: 'Of the 490 locos of the Aspinall 0-6-0 type . . .'
Page 225, 0-8-0 numbering	'12982–64' should read '12982–6'.
Page 228, line 5	Date should read '1938/9'.
Page 255, LMS No. 12493	This engine, while looking rather like a Furness 0-6-0 was actually built by the NBL Co. to a drawing of Manson's GSWR '361 class' — see *Volume 3, page 160*.
Page 261, line 5	The words 'Princess Royal' should have been printed between '4-6-2' and 'BR' on line 4 of the corrections list!

Further corrections, if notified or spotted, will be included in later volumes — please advise via the Publisher.

Additions to Volume Two

We have been taken to task for not including the ex-LNWR steam railmotors in *Volume Two* — vide our coverage of the NSR and LYR examples. However, the LNWR examples were regarded by the LMS as self-propelled carriage stock and numbered accordingly, rather than being regarded as locomotive stock. The difference was that they did not have the properly indepent engine portion which would have put them into the locomotive capital stock fleet. In any case, these vehicles were included in *An Illustrated History of LNWR Coaches* by one of the authors, published by OPC in 1978.

At the same time, we did omit to mention the 2ft. 6in. gauge engines belonging to the LNWR Engineering Department. These were used for permanent way construction on temporary tracks and when not in use were stored at Crewe. They bore neither LNWR nor LMS numbers but were named *Jim Crow*, *Platelayer* and *Kitchener*, dating from 1894, 1895 and 1914 respectively. All three lasted until 1941.

Additionally, as we had hoped would be the case, we were sent more details of the departmental Ramsbottom 'Special' tanks covered on *page 73*. From this information it would appear that No. 3323 was withdrawn in 1954, not 1959 and was outlived by CD3,6,7 and 8, not to mention *Earlestown*, all of which were withdrawn during 1957–9. No. CD7 was the very last survivor and may not have been broken up until 1960. The Wolverton engines of this type did carry their departmental numbers but the figures only appeared on the backs of the cabs.

Finally, Barry Lane has asked us to point out that the start of tender exchanges on the LYR 0-6-0s was not quite as precise as might be inferred from *page 212*. The procedure actually began after the Great War and became more noticeable during LMS days. He also makes the interesting supplementary comment that the rivets sometimes seen on the bunker sides of 2-4-2Ts — *e.g. Plates 397/8* — reflected LMS period repairs to previously flush-panelled bunkers.

As usual, we would like to thank our correspondents for their helpful comments and invite them to continue in the same vein if possible.

Index of Locomotive Classes (Capital Stock) covered in Volume Three

Listed in LMS number order

LMS NUMBER SERIES	PRE-GROUP COMPANY	WHEEL ARRANGEMENT	LOCOMOTIVE TYPE/CLASS		PAGE
14000	GSWR	2–4–0	J. Stirling '75 Class'		111
14001-14002	GSWR	2–4–0	Smellie '157 Class'		111
14010	CR	4–2–2	Neilson design		8
14100-14107	CR	4–4–0	Brittain design ('Oban Bogies')		9
14108-14115	CR	4–4–0	D. Drummond '80 Class'		10
14116-14137	GSWR	4–4–0	Smellie '119 Class'		112
14138-14156	GSWR	4–4–0	Smellie '153 Class'		114
14157-14202	GSWR	4–4–0	Manson '8 Class'		115
14203-14227	GSWR	4–4–0	Manson '336 Class'		120
14228-14243	GSWR	4–4–0	J. Stirling '6 Class'		122
14244-14270	GSWR	4–4–0	Manson '8 Class', '240 Class'		115
14271-14276	HR	4–4–0	Jones 'Strath Class'		169
14277	HR	4–4–0	Jones 'Skye Bogie'		169
14278	HR	4–4–0	Jones 'Bruce Class'		169
14279-14285	HR	4–4–0	Jones 'Skye Bogie'		169
14290-14310	CR	4–4–0	D. Drummond/Lambie '66 Class'		12
14311-14325	CR	4–4–0	McIntosh 'Dunalastair I Class'	⎫	14
14326-14336	CR	4–4–0	McIntosh 'Dunalastair II Class'	⎬ saturated	16
14337-14348	CR	4–4–0	McIntosh 'Dunalastair III Class'	⎪	20
14349-14365	CR	4–4–0	McIntosh 'Dunalastair IV Class'	⎭	23
14366-14378	GSWR	4–4–0	Manson '18 Class'		125
14379-14396	HR	4–4–0	Jones 'Loch Class'		172
14397-14416	HR	4–4–0	P. Drummond 'Small Ben Class'		176

LMS NUMBER SERIES	PRE-GROUP COMPANY	WHEEL ARRANGEMENT	LOCOMOTIVE TYPE/CLASS	PAGE
14417-14422	HR	4-4-0	P. Drummond 'Large Ben Class'	179
14430-14433	CR	4-4-0	McIntosh 'Dunalastair II Class'	16
14434-14437	CR	4-4-0	McIntosh 'Dunalastair III Class' superheated	20
14438-14460	CR	4-4-0	McIntosh 'Dunalastair IV Class'	23
14461-14508	CR	4-4-0	Pickersgill '113 Class'	27
14509	GSWR	4-4-0	Whitelegg rebuild of Manson 4-cylinder design	126
14510-14521	GSWR	4-4-0	P. Drummond '131 Class', '137 Class'	127
14522-14523	HR	4-4-0	Cumming design	182
14600-14608	CR	4-6-0	McIntosh '55 Class'	32
14609-14618	CR	4-6-0	McIntosh '908 Class'	33
14619-14626	CR	4-6-0	Pickersgill '191 Class'	39
14630-14655	CR	4-6-0	Pickersgill '60 Class' (some built by LMS)	41
14656-14672	GSWR	4-6-0	Manson '381 Class'	130
14673-14674	GSWR	4-6-0	Manson '128 Class'	133
14675-14693	HR	4-6-0	P. Drummond 'Castle Class'	183
14750-14751	CR	4-6-0	McIntosh '49 Class'	35
14752-14755	CR	4-6-0	McIntosh '903 Class'	35
14756-14761	CR	4-6-0	Smith 'River Class' (designed for HR)	37
14762-14769	HR	4-6-0	Cumming 'Clan Class'	187
14800-14803	CR	4-6-0	Pickersgill '956 Class'	44
15000-15001	CR	0-4-2ST	D. Drummond '262 Class'	45
15010-15012	HR	4-4-0T	Jones design	189
15013-15017	HR	4-4-0T	Dübs & Co design	190
15020-15031	CR	4-4-0T	Lambie '1 Class'	46
15050-15054	HR	0-4-4T	Jones/Drummond designs	192
15100-15114	CR	0-4-4T	D. Drummond '171 Class'	47
15115-15124	CR	0-4-4T	Lambie '19 Class'	48
15125-15146	CR	0-4-4T	McIntosh '92 Class'	50
15147-15158	CR	0-4-4T	McIntosh '104 (Balerno) Class'	53
15159-15240	CR	0-4-4T	McIntosh '439 Class'	55
15241-15244	GSWR	0-4-4T	Stirling '1 Class'	134
15245-15254	GSWR	0-4-4T	Manson '326 Class'	134
15260-15269	GR	0-4-4T	McIntosh '439 Class' (built by LMS)	55
15300-15307	HR	0-6-4T	P. Drummond design	195
15350-15361	CR	4-6-2T	Pickersgill '944 Class'	61
15400-15405	GSWR	4-6-4T	Whitelegg '540 Class'	136
16000	CR	0-4-0ST	Barclay design	65
16001-16002	CR	0-4-0ST	Dübs '446 Class'	65
16003-16007	CR	0-4-0ST	Neilson '502 Class'	65
16008-16039	CR	0-4-0ST	D. Drummond/McIntosh '264 Class'	67
16040-16043	GSWR	0-4-0ST	Various designs	139
16044-16049	GSWR	0-4-0T	Manson '272 Class'	141
16050-16051	G&PJt	0-4-0ST	Neilson design	141
16080-16085	GSWR	0-4-4T	Manson '266 Class'	142
16100-16102	CR	0-6-0ST	D. Drummond '272 Class'	70
16103-16117	GSWR	0-6-0T	Manson '14 Class'	143
16118-16119	HR	0-6-0T	Stroudley design	196
16150	CR	0-6-0ST	Brittain '486 Class'	71
16151-16173	CR	0-6-0T	McIntosh '498 Class'	74
16200-16224	CR	0-6-0ST	D. Drummond '323 Class'	71
16225-16229	CR	0-6-0ST	Lambie '211 Class'	73
16230-16376	CR	0-6-0T	McIntosh '29 and 782 Classes'	77
16377-16379	GSWR	0-6-0T	P. Drummond '5 Class'	146
16380-16382	HR	0-6-0T	P. Drummond design	198
16383	HR	0-6-0T	Stroudley design (as per 16118-19) — wrongly numbered after 16380-2 and the mistake never rectified	196
16400-16409	GSWR	0-6-2T	Whitelegg '1 Class'	147
16410-16427	GSWR	0-6-2T	P. Drummond '45 Class'	147
16500-16505	CR	0-8-0T	McIntosh '492 Class'	82
17000-17020	CR	0-4-2	Brittain '670 Class'	85
17021-17022	GSWR	0-4-2	P. Stirling '141 Class'	149
17023-17027	GSWR	0-4-2	J. Stirling '208 Class'	149
17028-17034	GSWR	0-4-2	J. Stirling '187 Class'	149
17035-17075	GSWR	0-4-2	J. Stirling '221 Class'	151
17100	GSWR	0-6-0	P. Stirling '58 Class'	153
17101-17102	CR	0-6-0	Ex-Solway Junction Railway engines	87
17103-17111	GSWR	0-6-0	J. Stirling '13 Class'	154
17112-17164	GSWR	0-6-0	Smellie '22 Class'	154
17165-17184	GSWR	0-6-0	Smellie/Manson '306 Class'	157
17185-17202	GSWR	0-6-0	Manson '160 Class'	158
17203-17212	GSWR	0-6-0	Smellie '22 Class' rebuilt by Manson	154
17230-17473	CR	0-6-0	D. Drummond '294 Class' ("Jumbo") plus Lambie/McIntosh continuation	87
17474-17507	GSWR	0-6-0	Manson '361 Class'	160
17508-17524	GSWR	0-6-0	Manson '17 Class'	163
17550-17645	CR	0-6-0	McIntosh '812 Class'	95
17646-17649	CR	0-6-0	McIntosh '30 Class'	99
17650-17692	CR	0-6-0	Pickersgill '300 Class'	100
17693-17704	HR	0-6-0	P. Drummond design	199
17750-17764	GSWR	0-6-0	P. Drummond '279 Class'	166
17800-17804	CR	2-6-0	McIntosh '34 Class'	102
17820-17830	GSWR	2-6-0	P. Drummond '16 Class'	167
17900-17904	CR	4-6-0	McIntosh '918 Class'	103
17905-17915	CR	4-6-0	McIntosh '179 Class'	104
17916-17930	HR	4-6-0	Jones 'Goods'	203
17950-17957	HR	4-6-0	Cumming 'Clan Goods'	205
17990-17997	CR	0-8-0	McIntosh '600 Class'	106